Between Democracy and Technocracy

Between Democracy and Technocracy

Regulating Administrative Guidance in Japan

Franklin Barr Lebo

LEXINGTON BOOKS
Lanham • Boulder • New York • London

Published by Lexington Books
An imprint of The Rowman & Littlefield Publishing Group, Inc.
4501 Forbes Boulevard, Suite 200, Lanham, Maryland 20706
www.rowman.com

Unit A, Whitacre Mews, 26-34 Stannary Street, London SE11 4AB, United Kingdom

British Library Cataloguing in Publication Information Available

Library of Congress Cataloging-in-Publication Data

Includes bibliographic references and index.
ISBN 9781498562218 (cloth : alk. paper)
ISBN 9781498562225 (electronic)

∞ ™ The paper used in this publication meets the minimum requirements of American National Standard for Information Sciences Permanence of Paper for Printed Library Materials, ANSI/NISO Z39.48-1992.

Printed in the United States of America

Dedication

I dedicate this work to my wife and soulmate, Dr. Amy Eloise Lebo, for her love, encouragement, and support. As we have just welcomed our first child, Elora Eloise Lebo into this world, my heart is filled with hope in our future and gratitude to be your lifelong friend.

Contents

List of Illustrations

List of Tables

Acknowledgments

I gratefully acknowledge the ongoing support of many individuals and institutions whose assistance made this project possible.

First, my great thanks to the Mansfield Fellowship Program (MFP) whose representatives assisted me in numerous respects from arranging my introductions to the National Personnel Authority (NPA) in Japan, to contacting the Mansfield Fellows through mass e-mails, to completing interviews with me at the Mansfield Foundation Office in Tokyo, to providing me with additional documentary data about the MFP, and inviting me to attend a Mansfield Fellowship briefing at the National Graduate Institute for Policy Studies (GRIPS) in Tokyo on the legislative process by then current Fellow, Rhonda Johnson. Specifically, my tremendous thanks to Director Paige-Cottingham Streater, Esq., Ms. Sara Seavey, Ms. Teruyo Kuramoto, and Ms. Chie Igarashi.

Second, my deepest thanks to the following Mansfield Fellowship alumni for agreeing to participate in this research project providing reams of data, lengthy interviews, responding to numerous follow up e-mails, surveys, and phone calls including: Mr. Jay Biggs, Mr. David Boling, Mr. Leo Bosner, Mrs. Yukiko Ellis, Mr. Ken Ishimaru, Mr. Jim Kariya, Dr. Ken Kobayashi, Mr. Brent Maier, Esq., Mr. Michael Panzera, Esq., Mr. Naveen C. Rao, Esq., Mr. Jeffrey E. Seay, Mr. Larry Swink, and Mr. Anthony J. Waller. Mr. Biggs also went out of his way to guest lecture to my Asian politics students at Kent State University on the Japanese administrative system and inspired them with his experiences as a Mansfield Fellow. I also thankfully acknowledge additional Mansfield Fellows who asked to remain anonymous and assisted in this research project only with the assurance that strict confidentiality as to their identities would be maintained as a matter of national security and international diplomacy. My thanks as well to Michael J. Marcus and James J. Miller for posting their Mansfield Final Reports online for public review.

Third, I respectfully acknowledge Japan's NPA for their thoughtful assistance with special appreciation to Dr. Tadakazu Watanabe, then serving director of the International Affairs Division, and Ms. Kazuko Kamada, international affairs officer, for taking time from their busy schedules to interview with me at the NPA headquarters in Kasumigaseki, Tokyo, and fulfilling my numerous requests for data. In addition, my thanks to Dr. Fredrick Butcher for his timely and excellent translation assistance of various NPA-supplied documents.

Fourth, my sincere thanks to the outstanding anonymous reviewers who provided insightful edits, inquiries, and recommendations.

Finally, my thanks to the wonderful staff at Rowman & Littlefield for making this publication possible.

Any errors, omissions, and opinions expressed in this work are my responsibility alone.

Introduction

LINKING DEMOCRACY AND BUREAUCRACY

Democracy is on the rise worldwide as we are in the throes of Samuel P. Huntington's (1991) fourth wave of democracy building. In fact, we may be witnessing a fifth wave as democracy has gained in popularity (Mueller 1999) with efforts to establish democratic regimes in the Middle East, such as in Iraq, Afghanistan, Egypt, and Libya. With these efforts has come significant theorizing about how to encourage democracy in states without a tradition of democratic rule and to do so within a relatively short timeframe (see, e.g., Esman 1999; Rondinelli and Montgomery 2005; Hendriks 2006; Honig 2007; Hauss 2009; Hook 2010). Democratic regime theorists recognize that the answer does not only lie in establishing legislatures, executives, and judiciaries along with competent security forces to keep the peace. Rather, democracy building also requires strong administrative institutions staffed by professionals to insulate states from capture or recapture by oppressive regimes (Rondinelli and Montgomery 2005).

No modern democracy—no matter how participatory—depends exclusively on elected officials, the rule of law, and security forces to govern. Rather, some decision-making authority must be delegated to administrative agents. Someone must count the ballots. Someone must collect taxes. A state where the United Nations must perpetually intervene to ensure elections are free and fair cannot guarantee that governance will continue after fingers stained in purple ink have been scrubbed clean. As Lupia and McCubbins (2000, 304) intone, "[d]elegation is a necessary component of modern democracy." As such, a bureaucracy responsive to elected officials—and thereby to "the people"—is an essential ingredient of deliberative democracy.

Thus, to build and sustain a lasting democracy requires careful thinking about the middle ground between democracy and technocracy. Huntington provides a helpful starting point for this inquiry by offering a parsimonious definition of "democracy." As he explains, "[t]he central procedure of democracy is the selection of leaders through competitive elections by the people they govern" (Huntington 1991, 6). Similarly, classic definitions of democracy emphasize the participation of the citizenry in the government, contestation for political office, and elections that are free and fair (Dahl 1989).

Although elegantly simple measures of whether a state is a democracy, in practice the implications of this minimalist definition are profound and complex. First, this definition forgives democracies for many defects and deficiencies. As Huntington (1991, 10) clarifies,

> [g]overnments produced by elections may be inefficient, corrupt, shortsighted, irresponsible, dominated by special interests, and incapable of adopting policies demanded by the public good. These qualities make such governments undesirable but they do not make them undemocratic.

John Mueller (2010, 46) puts it even more bluntly that democracy is merely a "superior gimmick" characterized by decision-making that "is often muddled, incoherent, and slow, [with] . . . results [that] are sometimes exasperatingly foolish, shortsighted, irrational, and incoherent." Terry M. Moe (1989, 267) expands the list of maladies to include "inertia, parochialism, fragmentation, and imperialism." In response, Robert Dahl (1989) might shrug that such is the nature of ordinary polyarchy whenever power is shared amongst a plurality of interests. Yet, all four thinkers would seem to concur that if the essential decisions of government are made by elected officials, so long as the electoral process was free, contested, and fair—or perhaps even intermittent, random, and perverse (Mueller 2010, 47)—it is by definition a "democracy."

Second, the significance of this definition of democracy is that it brings into sharp relief the links between a healthy democracy and its technocratic elites. The first key point is that administrative deference to elected bodies is the expected norm in a democracy. As the giant of administrative studies, Max Weber, clarifies, "[t]he individual bureaucrat cannot squirm out of the apparatus in which he is harnessed . . . [but rather] is only a single cog in an ever moving mechanism which prescribes to him an essentially fixed route or march" (Weber 2003 [1946], 8). This expected deference is thus essential to the relationship between administrators and elected officials.

In addition to administrative deference, the corollary is that administrators will avoid making critical decisions without guidance from elected officials. In other words, elected officials express their interests through cues such as legislation and administrators exercise restraint in carrying out those directives. Thus, if civil servants rely on their own decision-making, it will be in a limited and objectively "responsible" manner. Their actions are dictated by the relevant policy problem, rather than individual level motivations and preferences such as jurisdictional power (King 1997).

Bureaucracy and Democracy in Japan

An excellent starting point for examining this often rocky relationship between democracy and the administrators who make it work is Japan which experienced democracy in fits and starts beginning in 1868 with the start of constitutional government. Since World War II, the Japanese people have remained ever cognizant that democracy is often more unstable and fragile than it seems. As Huntington (1991, 10) clarifies,

> conceivably a society could choose its political leaders through democratic means, but these leaders might not exercise real power. . . . If those democratically elected decision makers become . . . simply a façade for the exercise of much greater power by a nondemocratically chosen group, then clearly that political system is not democratic. Legitimate questions may be raised, for instance, as to whether the elected governments in Japan in the late 1920s . . . were sufficiently dominated by their military as not to be truly democratic.[1]

Huntington's observation is thus particularly apt as this project explores the link between democracy and bureaucracy by examining Japan.

The modern Japanese system of public administration was established by the United States after World War II. As will be explained in greater depth in chapter 1, Japan is home to a structurally "typical" administrative apparatus in a parliamentary system vis-à-vis other advanced democratic states. It has multiple agencies called "ministries" organized around various policy areas with relatively clear goals and functional lines. For instance, the Ministry of Internal Affairs and Communications (MIC) crafts communications policy (see chapter 2), the Ministry of Foreign Affairs (MOFA) handles foreign policy (see chapter 3), the Ministry of Land, Industry, Transportation, and Tourism (MLITT) is responsible for those four broad policy areas (see chapter 4), and the Ministry of Health, Labor, and Welfare (MHLW) presides over topics from healthcare to working hours for laborers (see chapter 5).

Whether in good or bad times, praise or blame for the state's economic health and international standing is invariably laid at the doorstep of Japan's administrative elite. In fact, in the decades following World War II, the Japanese government enjoyed a positive reputation for getting the job done. Not only had it presided over "miracle growth," but also efficiently managed the oil shocks of the 1970s and quickly balanced its books in the early 1980s (Scheiner and Muramatsu 2009, 181). As a result, "administrative guidance" meant to the average citizen "being able to recognize problems in their infancy and develop solutions before problems spiraled out of control" (Scheiner and Muramatsu 2009, 181).

Yet, with the downturn of the economy since the late 1980s turning into the financial crises of the 1990s (Nakamura 2012), the bureaucracy became charged with standing in the way of dynamic readjustment to changing international circumstances as China rose again.[2] As a result, the bureaucracy has been the target of tremendous criticism and sustained reform efforts, with the dust barely settled on the most recent changes in 2014 (Mishima 2017, 9). The logic of "decentralization and deregulation" underlying nearly all of these reforms is the perception that the discretionary powers vested in administrative elites must be curtailed. This was in order to enhance the power of Japan's elected representatives to ensure their policy preferences are implemented (Suleiman 2003, 165–68, 245). Some have even asserted as early as the mid-1990s this means a "Blueprint for a New Japan" (Ozawa 1994).[3] Other democracies in East Asia from Taiwan to South Korea also saw significant administrative reform in the name of democracy as the Asian Financial Crisis swept the region beginning in 1997 (Ehrhardt 2009).

In turn, understanding these attempts at administrative reform in Japan and East Asian democracies more broadly prompted political scientists to turn to principal agent (PA) theory. In so doing, scholars have discovered that when properly adjusted for the historical context of Asian democratic history along with the existence of multiple principals in Asian policymaking, the relationship between bureaucracy and democracy has many lessons to teach.

Connecting Principal Agent Theory to Japan

Since PA theory provides the foundation for this text, this chapter begins by quickly introducing PA theory and then extends PA theory building off of work by George Ehrhardt (2009) and others who address Japan and East Asia more broadly. Finally, this conversation serves as a bridge to the broader questions that need to be addressed.

Principal Agent Theory

PA theory is familiar to many fields of inquiry including students of classical contract law. A "principal" is "[o]ne who authorizes another to act on his or her behalf" (Garner 1996, 499). The designated individual is the "agent" meaning "[a] person who is *authorized* to act for or in place of another" (Garner 1996, 23, emphasis added). In forming this contractual relationship, the parties have established an "agency" which is "[a] fiduciary relationship created by [an] express or implied contract or by law, whereby one party (the agent) may act on behalf of another party (the principal) and bind that other party by words and actions" (Garner 1996, 23). As this clarifies, duties between principals and agents are founded in the *authority* to act and cemented by mutually entrusted obligations. As Nachmias and Rosembloom (1980, 3) clarify, the distinction between power and authority describes the nature of this relationship in that power may involve force whereas authority involves voluntary compliance by those who choose to suspend their own judgment and follow orders. Authority then breaks down into "ideal types" as Max Weber (1969, 7) explains with modern society evidencing "rational-legal" authority that in "its pure type is best presented by bureaucracy . . . where obedience is not owed by anybody personally but to enacted rules and regulations." Once the principal clearly delegates authority, then, the agent must act with care to ensure that the interests of the principal are served *within the scope of the authority granted.*

When agents fail to do so, a democracy becomes run by an unaccountable "tyranny of experts." Thereupon, the link between democracy and bureaucracy has been broken (Lupia and McCubbins 2000, 292). In that case, significant "*agency loss*" or "drift" occurs since "the behavior of agents leaves principals [not simply] unable to achieve their goals in an efficient manner [, but unable to] realize them at all" (Gormley and Balla 2008, 59). When this happens, democracy and bureaucracy have been decoupled as political principals have *abdicated* their authority to *unresponsive* administrators. Thus, this is when delegation becomes abdication (Lupia and McCubbins 2000, 295, emphasis in original).

In such instances, administrators are not exercising their discretion to make routine procedural decisions to carry out their duties, but instead are actively pursuing their own preferences which may or may not conflict with those of political principals (Meier and O'Toole 2006). Thus, "discretion" meaning *administrative decision-making absent directives from political principals* is not in itself destructive of democracy. Rather, the purpose for which that discretion is exercised is the key.

In a democracy, then, a principal-agent relationship exists not only between the citizenry and elected officials, but also between elected officials and appointed government administrators. Agents' powers are thus clearly articulated and agents under the rational-legal ideal type are expected not to exceed the bounds of their authority set by principals. Examining the link between democracy and bureaucracy is therefore equivalent to examining this principal-agent relationship. In doing so, the question is not whether administrators are failing in the technical aspects of their duties to the state. Again, the focus is not on whether a bureaucracy is performing efficiently or inefficiently since a democracy may accommodate poor performance. Rather, the focus is on bureaucratic drift manifested through the exercise of agents' discretion.[4]

In summary, a democracy exhibits pluralism in policy *formulation*, manifests chains of delegated authority from political principals to bureaucratic agents in policy *implementation*, and minimizes bureaucratic drift whereby bureaucrats pursue their own discretionary preferences in the absence of directives from principals. In so doing, a democracy ensures *responsiveness* in public administration as principals retain the power to determine policy objectives. Conversely, where a political system does not exhibit pluralism in policy formulation, implements policy through a hierarchical model emphasizing efficiency rather than chains of delegation, and/or demonstrates severe bureaucratic drift with agents consistently pursuing their own policy preferences, either political principals must curtail bureaucratic drift or the state will cease to be a democracy.

Principal Agent Theory in East Asia

Fitting PA theory to the experience of East Asian democracies has involved significant trial and error along with careful rethinking (e.g., McCubbins and Noble 1995; Nakano 1998; Ginsburg 2001; Ramseyer and Rasmusen 2001; Miller 2005; Baum 2005; Baum 2007; Yamamoto 2008; Ehrhardt 2009). The first theoretical challenge for PA theory was accounting for the disparate historical context of East Asian democracies. Specifically, East Asia is remarkable for democracies such as Taiwan, South Korea, and Japan that have experienced protracted periods without significant inter-party competition. This has resulted in cozy relationships between ruling parties and their bureaucracies (Ehrhardt 2009). As a result, "formal channels of control and accountability were less important" for ruling parties which often became coopted by administrative elites (Ehrhardt 2009).

Turning specifically to Japan, Scheiner and Muramatsu (2009, 182) first point out that the relationship between bureaucrats and politicians is generally considered to be competitive and thus a zero-sum game in the political science literature (see McCubbins et al. 1989). However, due to the long cozy relationship between the Liberal Democratic Party (LDP) and bureaucracy in Japan in the post-war era, a *positive-sum* relationship developed. Indeed, until 1993, powerful iron triangles between the LDP, bureaucrats, and business were frequently observed in Japan (McElwain, Mori, and Reed 2009). These cozy relationships allowed the participants to often achieve their policy preferences through cooperative rather than competitive negotiations.

Second, responding to McCubbins et al. (1989) assertion that it is feasible for principals to develop methods to constrain bureaucratic behavior, Scheiner and Muramatsu (2009) point out that in the absence of a competitive relationship due to the LDP's long tenure and the existence of said iron triangles in Japan, political principals never developed those methods. As they assert,

> [s]pecifically speaking about Japan, we argue that single-party dominance—the LDP ruled Japan without interruption from 1955–1993—led to an unusually stable and close relationship between the ruling party and central bureaucracy. As a result, Japan's parliament (Diet) never established structures that would allow it to obtain policy information on its own and develop legislative capacity in a number of policy areas . . . Instead, single-party dominance promoted great bureaucratic discretion . . . (Scheiner and Muramatsu 2009, 183).

Scheiner and Muramatsu (2009) then illustrate this point through an evaluation of the banking sector and the deterioration of iron triangles due to the financial crises of the 1990s when the bureaucracy worked with the eight party coalition in 1993–1994 to the chagrin of the LDP.

However, they do not finish their powerful insight by making the broader structural point. In short, this cozy relationship between the LDP and bureaucracy along with the positive-sum game prior to 1993, meant that the legislature never created the structures to allow the National Personnel Authority (NPA) (see chapter 1) to effectively fulfill its mission to constrain bureaucratic drift and thus limit the exercise of bureaucratic discretion. Yes, the formal structures were never developed by the Diet to constrain bureaucratic drift. More accurately, however, the formal structures were never developed by *the Diet* which would have allowed *the NPA*—until 2014, the legislatively designated agency—to control bureaucratic drift thereby fulfilling its structural purpose.[5]

The second issue for PA theory was its emphasis on unified actors (i.e., politicians and the bureaucracy), without accounting for multiple principals with conflicting interests. Yet, these conceptual difficulties are not insurmountable. As Ehrhardt (2009, 629) explains, "single principals will exercise control over the bureaucracy to pursue their electoral interest; multiple principals, on the other hand, will constrain political control over the bureaucracy to prevent intervention by rivals."

These elegantly simple observations lead Ehrhardt (2009, 630) to an equally parsimonious conclusion that the presence of multiple principals drives administrative reform. How is this possible? Multiple principals compete with each other which creates the necessary breathing room for bureaucrats to innovate. This might involve intentionally pitting political principals against each other, but may also be "unintended, as when bureaucrats genuinely don't know which principal's goals to pursue" (Ehrhardt 2009, 630). Over time, this bureaucratic "drift" often engenders a backlash against administrators since principals become frustrated with a bureaucracy that seems to be pursuing its own agenda. As a result, principals misdirect their frustration at their own impotence of calling for reforms or curtailing the bureaucracy itself. To push Ehrhardt to the logical conclusion, the appropriate response would instead be to address whatever underlying substantive issue was originally on the table while enhancing institutional mechanisms for clearer messaging to administrators. In other words, let administrators know the difference between policy wants and needs. Then give them the resources to get the job done.

Ehrhardt has struck the equivalent of theoretical gold as he then explains how Japanese administrative reform in the 1990s tracks this cycle. Led by a fractured and fractious LDP coalition, political principals constrained each other's power permitting bureaucratic drift in policy affairs. As a result, agents wielded their discretion to pursue their own preferences. The result was increasingly irritated factions of principals who ultimately united to enact in 1999 the most significant overhaul of Japan's administrative system since World War II (Ehrhardt 2009, 632–35).[6] The consequences for Japan's administrative system in recent years have been momentous.

Turning to the 2000s, following three lackluster and short lived prime ministers after Junichiro Koizumi's unexpectedly long tenure (2001–2006), the LDP lost control to the reformist Democratic Party of Japan (DPJ) in 2009. The DPJ wasted no time under Yukio Hatoyama in significantly reducing bureaucratic drift by eliminating important channels by which the bureaucrats had come to influence policy formulation (Nakamura 2012, see chapter 1 for more details on the evolution of the relationship between bureaucrats and politicians).

Meanwhile, the power of the prime minister to make decisions without consulting with either his party or the ministries was significantly enhanced, a trend first traceable to the agenda of Morihiro Hosokawa in the mid-1990s (see Inoguchi 2011; Shinoda 2011; Noble 2011; Krauss and Nyblade 2005; Takayasu, 2005; Mulgan 2003). Despite the poor performance of the DPJ to reinvigorate Japan's economy under the new structure resulting in declining public trust, Japan nevertheless seemed poised under the DPJ to continue this consolidation of executive power (Nakamura 2012). However, this changed when the triple disasters of the earthquake, tsunami, and meltdown of the Fukushima nuclear disaster struck in March of 2011. In particular, Prime Minister Naoto Kan (June 2010–August 2011) was unable to effectively coordinate disaster management and relief due to his disdain for bureaucratic involvement in policy formulation. This resulted in significant confusion, revealed the lack of expertise of politicians to manage complex policy problems, sharply decreased the trust in the central government from 18.50 percent in January to 12.45 percent in April, and resulted in Kan's resignation after just 452 days in office (Nakamura 2012, 254, *passim*). His successor, Prime Minister Yoshihiko Noda (2011–2012), then sought to restore the involvement of the former channels of influence of the bureaucracy and the party (see chapter 2). This trend was continued with Shinzo Abe (2012–present), the most powerful "Teflon premier" since Junichiro Koizumi (2001–2006) and a voice for hawkish conservativism (Tanikawa 2017). As Akira Nakamura (2012, 256) succinctly concludes, "[i]n the final analysis, there is a pendulum swing: too much dependence on central bureaucracy can generate problems, while disregarding the central agencies and professionals can also impede a reliable democratic system of government."

While there is indeed a pendulum, the centralization of power in the executive has swung too far. In the Shinzo Abe era, Japanese bureaucrats need greater authority and resources, but with heightened independent monitoring to reduce bureaucratic drift. This recognizes the trend in modern Japan and democracies more generally away from representative government in the classical sense and toward responsive or "monitory" democracy—"the politics of mutual monitoring of government and citizens evolving around transparency and accountability" (Inoguchi and Purnendra 2011, 6). Likewise, this outcome accords with the literature on modern politics in Japan that formal structures to reduce bureaucratic drift were never developed due to the long tenure of the LDP and its close ties to the bureaucracy. These structures are needed today and have begun to appear with the formation of the Cabinet Bureau of Personnel Affairs (CBPA) with a direct report to the prime minister. A number of the NPA's traditional functions have been handed over to the CBPA in this new model such as the "legal authority to direct ministries regarding recruitment, promotion, training and career development, and performance" review (Mishima 2017, 9).

Despite some movement in this direction, this assertion is admittedly controversial and diametrically opposed to the thrust of Japan's modern reform efforts largely aimed at *curtailing bureaucratic authority, enhancing executive power in bureaucratic affairs, and eliminating independent monitoring*. However, rather than struggling against Japan's predilection for administrative governance or what some call the "Asian Model" (Nakamura 2012), these tendencies ought to be accepted if not embraced by political principals (i.e., the Diet or legislature). In so doing, the exercise of bureaucratic expert administrative authority will become the accepted *raison d'etre* of Japan's system of public administration. At the same time, despite the cost, oversight of the ministries must either be forcefully exercised by elected officials or delegated to a powerful agency to ensure that bureaucratic drift is minimized. The alternative is the present system of increasing meddling by political principals (i.e., the Prime Minister and Cabinet) acting through a weak monitoring agency (e.g., the NPA and/or the fledging CBPA) while the chains of delegation between the bureaucracy (ministries) and democracy (political principals) are increasingly stressed. This is so despite the enhanced powers of the Prime Minister and Cabinet in recent years. Absent this change in policy, Japan's administrative model will continue to underperform and when the pendulum swings away from the current powerful executive, we are very likely to once again see popular dissatisfaction and reform minded powerbrokers attack the bureaucracy as has happened numerous times in the past. Again, only by delegating *greater* authority to Japanese administrative agents, albeit supervised by a powerful intermediary such as the NPA or the CBPA, may bureaucratic drift be minimized and the exercise of discretion thereby contained. Thus, the following section provides a bit more insight into the research question and direction this inquiry will take.

Approaching the Study of Bureaucratic Drift in Japan

While Ehrhardt (2009, 631) is seeking to assess administrative reform in particular, this study evaluates bureaucratic drift as a more fundamental issue to the quality of Japanese democracy. In more theoretical speak, this inquiry addresses different dependent variables: namely, administrative reform and bureaucratic drift respectively. Specifically, Ehrhardt (2009, 631, emphasis removed) "argue[s] that the existence of multiple principals is the engine that drives administrative reform." Instead, this project addresses the deeper issue of bureaucratic drift noted by Ehrhardt (2009, 631): namely, *does competition by multiple principals create opportunities for bureaucratic drift*? The contribution of this study is in providing evidence directly from the ministries with primary examples on this question rather than through secondhand information. As is described below, the finding is that policy settings which include multiple principals (independent variable) are more likely to manifest bureaucratic drift (dependent variable). Policy setting excluding multiple principals (independent variable) are more likely to manifest less bureaucratic drift (dependent variable). The effect of the independent variable on the dependent variable is equally important to underscore. Specifically, *variation in agent discretion is the critical effect of the independent variable (i.e., number of principals) on the dependent variable (bureaucratic drift).* Once again, *"discretion" is "administrative decision-making absent directives from political principals."*

This work contributes to the extant literature in a number of important respects. First, this study looks at administrators in the ministries themselves holding up specific instances of policy choices and decisions to illustrate the points made. This is quite different from the mostly armchair or arm's length studies that typically review the Japanese administrative system. Second, this work proposes a different dependent variable in that most studies are focused on administrative reform whereas the focus here is on bureaucratic drift. Third, this project discusses the important *effect* of bureaucratic discretion. Fourth, while salient to the quality of the Japanese democracy in particular, these findings may be leveraged to a larger conversation about the relationship between bureaucracy and democracy in the Asian context and perhaps beyond. Finally, this project proposes delegating *greater* authority to administrative agents, albeit supervised by a powerful intermediary, to minimize bureaucratic drift. Naturally, further research in other states is required to test the research question to determine its explanatory power in other settings.

Given this definition, evaluating the exercise of discretion of administrators is feasible if one's evidence is primarily from the administrators themselves. As presented in chapters 2 through 6, the data from the Mansfield Fellows and the extant literature on the ministries provides substantial grounds for assessing the effect (bureaucratic discretion) of the independent variable (i.e., the number of principals) on the dependent variable (i.e., bureaucratic drift). This is a relatively challenging prospect as penetrating Japan's government is quite complicated for a variety of reasons. Thus, few studies typically adopt this approach making this text hopefully a helpful addition to the literature.

With that goal in mind, let us stop to consider the structure of the book for a moment. First, we will turn in chapter 1 to briefly consider the structure of the Japanese government and the history and powers of the NPA which were curtailed and reassigned to the CBPA in 2014. While there are plenty of excellent texts that are referenced for much fuller histories of the Japanese system of public administration, what makes this rendition different is the inclusion of the NPA along with a comparative discussion of what was happening simultaneously in the United States and Japan from an administrative standpoint after World War II. In addition, this discussion will help to clarify why the Japanese system of public administration today, even after this most recent spate of reforms, lacks oversight which has contributed to bureaucratic drift.

To test these hypotheses, this inquiry adopts a research design based on a case study methodology. The case studies include five of Japan's ministries: the Ministry of Internal Affairs and Communications (MIC) (chapter 2), the Ministry of Foreign Affairs (MOFA) (chapter 3), the Ministry of Land, Industry, Transportation, and Tourism (MLITT) (chapter 4), Ministry of Health, Labor, and Welfare (MHLW) (chapter 5), and the Ministry of Economy Trade and Industry (METI) (chapter 6).

In response to criticism that case studies cannot aid in theory building, there are multiple rationales for adopting this approach. One particularly apt for this book is the "revelatory" case rationale where research is conducted into previously inaccessible social phenomena (Yin 2003, 42–43). To be clear, the first challenge when seeking to evaluate Japan's administrative system is the problem of securing access. As Kenneth Goldstein (2002, 669) explains, "[f]rankly, 'getting the interview' is more art than science and, with few exceptions, political scientists are not particularly well known for our skill at the art of 'cold calling.'" Administrative agents are busy elites with few incentives to assist in research inquiries. As Neil Stephens (2002, 205) explains, "[w]hether elite is defined in terms of social position relative to the researcher conducting the interview . . . or relative to the average citizen in society, they are still clearly in a position of power and raised social stature."

The initial task for this inquiry, therefore, was to identify access points and then to establish credibility, legitimacy, and the trust necessary to engage bureaucratic elites in an in-depth qualitative analysis. In this vein, the author determined that access to the NPA of Japan was essential. As the official oversight agency for personnel issues in Japan, the NPA is the only governmental body with ongoing systemic connections to the entire Japanese administrative system (see chapter 1). As such, it is the only body with connections to a cadre of bureaucratic elites at Japan's various ministries. As to the "elite" status of NPA administrators, the NPA reports directly to the Japanese prime minister on an annual basis so that political principals in the Diet (legislature) may act directly on its recommendations to change the personnel system through new legislation. As such, NPA personnel are in a position of power and raised social status compared to both the researcher and general public and are thus "elites." With this access point identified, the next challenge was to ascertain an effective entrée into the NPA's extensive network. Additionally, since NPA personnel rotate out of their positions every two years, locating a suitable subset of administrators associated with the NPA to target also proved challenging.[7]

The answer proved to be an NPA facilitated personnel "exchange" program designed to build a cadre of experts deeply embedded in the ministries. Specifically, the Maureen and Mike Mansfield Fellowship Program (described later) sends U.S. administrative personnel as participant-observers to Japan to better understand the decision-making styles employed in Japan's ministries. The goal is to ultimately enhance bilateral policy cooperation between the United States and Japan. During their fellowships in Japan's ministries, these personnel are managed both by the NPA and the Mansfield Foundation's Fellowship Program office in Tokyo. Thus, this inquiry required establishing relations with not one, but two sets of elites to complete this project.

Although the author first attempted to contact the NPA directly, this initial effort yielded no helpful results as requests for contacts and other communications went either unanswered or answered with automatic or formulaic replies. When this failed, the author approached the Mansfield Fellowship Program to gain access to both the requisite Mansfield Fellowship participants as interviewees and, ultimately, to the personnel in the International Affairs Division of the NPA.

Although this approach took several months, eventually it proved to be successful. Most importantly, Director Paige Cottingham-Streater of the Maureen and Mike Mansfield Fellowship Program agreed to assist in gaining access both to the Mansfield Fellowship alumni and to key NPA personnel in the International Affairs Division. As to Mansfield Fellowship alumni, blast e-mails were sent out to the fellows from the Washington, DC, headquarters which helped to generate interest amongst the Mansfield alumni. As to requisite NPA personnel, access required a series of negotiations instigated by Cottingham-Streater with the NPA, which ultimately culminated in formal introductions at the NPA headquarters in Kasumigaseki, Tokyo.

Without relating the many months of scheduling, planning, and negotiating that this process entailed, after numerous telephone calls along with a formal recorded telephone interview with Cottingham-Streater along with e-mails and/or telephone calls with various staff members at the Washington, DC, and Tokyo, Japan, offices of the Mansfield Program, the author was finally ready to begin the interviews in Japan. The author then met the staff of the Tokyo Office of the Mansfield Fellowship Program first in person including Cottingham-Streater, Ms. Teruyo Kuramoto, and Ms. Chie Igarashi. Formal interviews were then conducted with these three interviewees and a variety of documents and other materials regarding the Mansfield Program were collected including the original news broadcast announcing the Mansfield Fellowship Program to the Japanese public, which included interviews with the first group of Fellows in Japanese. During this initial data collection and interview process, the relationship between the Mansfield Fellowship Program and the NPA along with the NPA's collaboration with various branches of the Japanese government became apparent. Indeed, each of the Mansfield Fellows must be placed by the NPA in a ministry for a year-long fellowship. In particular, the staff of the Tokyo Office of the Mansfield Program has garnered significant expertise since the start of the Mansfield program in 1994 liaising not only with the NPA, but also the various Japanese government ministries. Hosting foreign administrative personnel is a complex process that often required the joint efforts of the Mansfield Fellowship Program, the NPA, the sending U.S. federal agency, and the host ministry in Japan.

The second source of data for this project was gathered from the Mansfield Fellowship Program alumni. Since receiving its initial funding from Congress in 1994, the Maureen and Mike Mansfield Fellowship Program[8] has sent over 100 U.S. federal employees to participate in an exchange program with various levels of the Japanese government including ministries, courts, agencies, and even the Diet, Japan's national legislature (NPA 2006, 21). To enhance cooperation and friendly relations between the two nations while building a reservoir of bilateral policy expertise, U.S. federal employees undergo intensive training in Japanese politics, history, culture, and language for a year. They then travel to Japan for placement by the NPA for another year long stint in a ministry, agency, and/or other branch of relevant policymaking authority. To qualify for the program, the applicant must have served for at least two years immediately prior to entering the program as a federal employee, although those with three or more years of experience are preferred.

The process of selecting these federal employees is both competitive and stringently regulated including approval by the sponsoring U.S. agency and the presentation of a clear fellowship proposal before a panel of reviewers at the Mansfield Foundation clarifying how the proposed fellowship will be beneficial for bilateral relations. Once selected, the Mansfield Program and NPA undertake delicate negotiations and evaluations to determine an appropriate placement for the fellow in Japan with a ministry. This helps to ensure that the experience is rewarding (Watanabe 2011, personal communication).

At the same time, since Japanese administrators do not reciprocally travel to the United States, one might wonder why this program is an "exchange?" In short, the goals of the exchange program are ultimately diplomatic. In honor of the late Ambassador Mike Mansfield who believed that policy cooperation between the United States and Japan was essential, the Mansfield Program seeks to improve bilateral relations through administrative sharing. As such, the program expects a two way cooperative learning process to occur which allows the Mansfield Fellows to serve as human conduits for the bilateral transfer of knowledge. In this way, the program is a foreign policy tool of both the U.S. and Japanese governments and has generated long-term interagency enmeshment across the pacific.

Thus, as participatory observers, each Mansfield Fellow is expected to not only gain an insider's look into Japan's ministries, but also to lend Japanese administrators insight into how U.S. counterparts might approach similar policy problems. To arrange this experience, therefore, is complex. The director of the Mansfield Fellowship Program, personally participated in the lengthy negotiations for placement of every fellow in Japan since the first fellows in 1996. This process was quite similar to the process she undertook on behalf of the researcher when arranging for the researcher to meet with the NPA personnel in the International Affairs Division. However, for a full year placement, these negotiations take place on a much larger scale. As Cottingham-Streater (personal communication, October 22, 2009) explains of the negotiation process,

> With each Fellow's placement we start fresh. Some people have felt that after 15 years with a number of agencies that have participated, at this point it should be easy. But . . . for all 81 alums that have participated in the program, each negotiation for each placement office at each agency, ministry, and Diet office was a separate negotiation . . . because there were different people involved . . . [due to] the Japanese system of [personnel] rotation. So each time, we were reintroducing the program, its effectiveness, goals, and what [the Fellows] want to do each time.

In addition to the difficulty of arranging the requisite permissions and selecting the appropriate offices to host Fellows, ensuring that the projects selected for the Fellows would be meaningful is also a challenging task. She clarifies the following rationale underlying the initial set up of the Mansfield Fellowship experience,

> In those implementation details, we defined and described the kind of experience we wanted [the Fellows] to have. . . . [T]his is intended to be a practical experience. On-the-job training. Hands-on. Learning-by-doing. Getting an overall view of the Japanese government and how it works, the decision-making process. Building professional relationships. We thought [about] how we were going to do that if it were treated as though it were a program [where each participant has] independent research. . . .We wanted our program to be participative, engaged. So we made a point of being clear that the Fellows would be first of all accepted in their counterpart agencies because that was the goal of strengthening the relationship and having across the board [U.S.] federal expertise about Japan in a variety of agencies on a variety of issues. And we wanted to make sure that there was an opportunity for those post fellowship to stay engaged. . . . So it was important that they would be assigned to a counterpart agency. . . . It was important that they would be able to be involved on a daily basis in that Japanese workplace environment building relationships, getting to know the people.

Notably, this operational learning and sharing of collective knowledge occurs both informally and formally. Informally, Fellows directly interact with Japanese administrators in their host agencies. Formally, Mansfield Fellows work on designated projects, prepare a Final Report, complete lengthy administrator training programs at the NPA, and complete a debriefing session with the NPA before returning home (NPA 2006, 21). Finally, the Fellows must commit to continued service in a U.S. federal government post for two years after completing the Mansfield Fellowship Program.

As such, the nature and scope of the Fellowship for each individual participant must be carefully vetted and approved at multiple levels including significant involvement by the NPA. Thus, it is unsurprising, perhaps, that a single Mansfield Fellowship placement may take up to a year to arrange and Cottingham-Streater would spend about half of her year personally visiting various government offices to secure permissions for the fellowships that are described in chapters 2 through 6. Thus, given the uniqueness of this administrative exchange program, the Mansfield Fellowship Program alumni present a rare dataset thus lending additional support in a case study approach.

Despite the potential of plumbing this dataset, a practical question is where to begin? In short, the personal testaments and final reports prepared by Fellows, many of which are thickly descriptive, provided a starting place. As these reports are prepared during the fellowship placement or immediately after exiting Japan they are expressions of contemporaneous observations which is critical as memories fade since the first group of Mansfield Fellows traveled to Japan in 1996. Very little of this material is in the public domain and, in all but two cases, could only be obtained with the express permission of the Fellows themselves as neither the Mansfield Fellowship Program nor the NPA is at liberty to disclose these reports unilaterally. Yet, many of the MFP alumni have a variety of security, diplomatic, and professional concerns for releasing their final reports without redacting significant portions and still others required multiple assurances that the reports would not be shared with a larger audience without their express consent. Likewise, nearly all of the Fellows required that they be permitted to review the final case study reports on their personal experiences for accuracy, security, and diplomatic reasons. Thus, the author obtained a final review by most of the participants prior to submitting this project and greatly appreciates the additional nuances and corrections that the participants provided. This double-check thus enhanced the accuracy and validity of the details presented. At the same time, the validity of the reports themselves is to some degree enhanced as prior to leaving Japan, the Fellows must submit their Final Reports for review by the Mansfield Fellowship Program, their sending agencies in the United States, and the Japanese government. In addition, they must complete an exit interview with their Japanese hosts. Thus, while there certainly may be errors or omissions in the reports along with misinterpretations of events by both the Fellows and the researcher, there is at least a reasonable expectation of accuracy due to both the circumstances of the review and preparation of case studies based upon these reports and various checks that these reports have been subjected to in their initial creation by all salient parties.

Other primary data sources and materials were also consulted such as PowerPoint presentations delivered by the Fellows during their NPA personnel training seminars or for other inter and intra governmental presentations, contemporaneous reports and logs, reports prepared during the Fellowship for the particular Japanese agency, and subsequent reports and documentation prepared for their sending U.S. federal agency as a means of enhancing bilateral policy cooperation between the United States and Japan. As such, these data were gathered over two years through careful fieldwork. At the same time, some documentation necessary to complete this project was available in the public domain. In some instances, for example, Fellows published articles on their findings in scholarly journals or via the Mansfield Foundation's website and other publications. Likewise, in two instances, MFP alumni also posted their final reports online for public review. Lastly, the extant literature by area experts on the ministries is also drawn upon to bolster the analysis. Thus, altogether, sufficient triangulation of data for the five case studies presented in chapters 2 through 6 satisfies the requirement of construct validity.

Beyond collecting as much supporting documentary evidence as possible, two additional methodological safeguards were established to minimize the concern that inferences from the completed interviews might exhibit sample bias. First, this research took into account the need for sufficient "breadth" meaning that "[t]he more events one can describe, predict, and/or explain with a given proposition (*ceteris paribus*) the more powerful that proposition" (Gerring 2001, 103). As such, five of the possible ten different ministries that the NPA oversees[9] were selected to ensure sufficient breadth of the evaluation including the (1) Ministry of Internal Affairs and Communications (MIC), (2) Ministry of Foreign Affairs (MOFA), (3) Ministry of Land, Infrastructure, Transport, and Tourism (MLITT), (4) Ministry of Health, Labor, and Welfare (MHLW), and (5) Ministry of Economy, Trade, and Industry (METI).

Finally, the selection of the ministries themselves was consciously taken to avoid merely sampling those Mansfield fellows who have been placed with the most famous Japanese ministries. Japan's various ministries do not have equal reputations either for expertise or power (Johnson 1995; Hartcher 1998). Thus, the five ministries together represent a reasonable mixture of more or less popularly recognized ministries.

With this groundwork in place, chapters 2 through 6 demonstrate what "bureaucratic drift" looks like in action. To be more specific, five of Japan's ministries are brought into focus: the Ministry of Internal Affairs and Communications (MIC) (chapter 2), the Ministry of Foreign Affairs (MOFA) (chapter 3), the Ministry of Land, Industry, Transportation, and Tourism (MLITT) (chapter 4), and the Ministry of Health, Labor, and Welfare (MHLW) (chapter 5), and the Ministry of Economy Trade and Industry (METI) (chapter 6). Thus, chapters 2 through 6 present five case studies to assess whether *competition by multiple principals creates opportunities for bureaucratic drift*. Finally, the Conclusion considers the lessons to be drawn both from the design of the system and the case studies presented. In so doing, we hopefully will have a better sense of the events presently unfolding in Japan specifically and the dynamic relationship between democracy and bureaucracy on a global scale.

NOTES

1. Notably, Huntington's observations contain a deeply embedded normative assertion that a democracy is preferable to the alternative. As John Kingdon (43) observes, "[t]he model of a democratic government controlled by elected officials is not only our normative ideal, but also our dominant picture of empirical reality." Huntington's chosen example reinforces this observation since the rise of the military in the waning days of Japan's Taisho Era (1912–1926) led that state to a series of conflicts culminating in World War II. The implication is thus that this aggressive foreign policy was abetted by the decline of the democratic state in Japan and the rise of the military. Yet, even this seemingly clear case is difficult to call decisively as the

Japanese state had all the trappings of a modern democratic system from periodic elections to a constitution inked during the Meiji Era in 1868. Moreover, it was during its uncontested years as a democracy that it went to war with Russia in 1904 and China in 1894–1985. Thus, the implication that Japan's later aggressiveness was somehow an indication of a waning democratic state is questionable. Democratic Peace Theory is likely closer to the mark that democracies merely do not go to war *with each other*.

2. For instance, see Hartcher (1998), Johnson (1995), and Ozawa (1994).

3. This fire of anti-bureaucratic sentiment has frequently been stoked by a series of spectacular corruption scandals that have continually captured the attention of the media such as the discovery of 17,000 unreported offences by bureaucrats in 2003 (*Mainichi Daily News* 2003) along with cases of bribery, collusive bidding, and other acts of malfeasance since then, some of which even made their way into annual reports by government agencies (NPA 2010b, 71–72). Both popular media and the Japanese government are deeply enthralled by these salacious stories as they perceive a tangible and direct link between the value driven behavior of bureaucrats and efficient economic policy. As one article by the *Yomiuri Shimbun* (2008) in Japan explains on the topic of government reform, "[t]he development and securing of public servants with a high sense of ethics and an ability to cope flexibly with changes in situations both at home and abroad is an extremely important issue for Japan's future." This behavior has also been repeatedly linked to government excess, largesse, and inefficiency. As a result, the NPA was tasked with evaluating the effect of these scandals on the public's level of trust in government (Inatsugu 2011, 45–46).

4. Likewise, as the discussion of the National Personnel Authority (NPA) of Japan in this research project demonstrates, it is theoretically possible to *design* an *institution* to ensure delegation and prevent abdication in tune with McCubbins, Noll, and Weingast's (1987) expectations. The effectiveness of the NPA in accomplishing this mandate is, however, evaluated at length in chapter 1.

5. The responsibilities and powers of the NPA will be discussed in chapter 1 in the context of a historical review. Again, the NPA's *inability* to effectively restrain bureaucratic drift helps to explain why the ministries have exercised significant discretion as bureaucrats have pursued their own policy preferences.

6. See chapter 2 for more detail. Ehrhardt's specific case study of the principal-agent problem in the context of the Ministry of Foreign Affairs is introduced in chapter 3.

7. Ostensibly, the underlying rationale of this rotation system is to stave off potential corruption or biases in the system which might inhere if NPA personnel were to remain at their positions for an extended period of time. This has a long history in Japan reaching back to the time of the shoguns and even earlier in China where procedural approaches were oftentimes viewed as the most effective means to achieve substantive policy reforms. As the historian John Schrecker (1991, 32) maintains with regard to the Tang Dynasty (circa 618–907 AD),

> [T]here were specialized government ministries in the capitol and local officials who were part of a centralized hierarchy. There were regular evaluations of bureaucrats and rules of avoidance and rotation, which meant that no official could serve in his home area and none could hold an office for too long. This policy guarded against the tendency in the Han for officials to make government positions into private fiefs.

8. The program was established in honor of Mike Mansfield who was a U.S. Senator and ambassador to Japan (see, e.g., NPA 2006, 16).

9. Again, there are a total of eleven ministries in Japan, but the NPA does not oversee the Ministry of Defense. Even so, Mansfield Fellows have been placed at the Ministry of Defense as well, but have not been included in this project.

Chapter One

History, Powers, and Roles of the National Personnel Authority (NPA)

INTRODUCTION

For nearly sixty-five years, the National Personnel Authority (NPA) served as a unique actor in the Japanese system and a logical conduit for examining systemic questions related to public administration in Japan. To understand the history, powers, and roles of the NPA, however, requires first reviewing the structure of the Japanese government as a whole. We thus begin with the structure of the Japanese system including the evolution of the relationship between bureaucrats and politicians today. Second, the rather lively historical backdrop surrounding the establishment of the NPA is brought into focus. Third, the modern roles and powers that the NPA played and exercised in public administration are analyzed. Finally, this chapter notes the failure of the NPA to effectively accomplish its founding mission of constraining bureaucratic drift due to budget cuts, staffing constraints, and periodic political opposition to its continued existence. As a result, 2014 saw the emergence of a new entity to assume some of the core responsibilities of the NPA over policing the bureaucracy. Significantly, bringing the NPA or its successor, the Cabinet Bureau of Personnel Affairs (CBPA), into this conversation is certainly atypical in the available literature.

STRUCTURE OF THE JAPANESE GOVERNMENT[1]

The quote by Samuel Huntington (1991, 10) in the opening pages of this work is a surprisingly apt place to begin a discussion of the structure of the Japanese government. Once again, Huntington (1991, 10) explains that "conceivably a society could choose its political leaders through democratic means, but these leaders might not exercise real power." Compare this to the opening remarks that the legendary Chalmers Johnson (1995, 159) provides in his description of the Japanese system,

> All politically organized societies reveal a discrepancy between the formal and the actual, between the ideal and the normative functioning of their political institutions, and it is precisely the existence of this discrepancy that gives rise to political science . . . In the Japanese case . . . the expressed discrepancy is not merely a feature of interest to political scientists . . . [but] rather a structural feature around which the polity has been organized and evolved.

As this opening salvo implies, the question of where the actual locus of power is to be found in Japan is a matter of significant study and debate. In fact, five schools of thought have emerged since World War II which have provided very different answers to that question (Lebo 1999). The following outlines these perspectives while explaining the structure of the Japanese government. It concludes by outlining recent trends in the evolution between the role of politicians and bureaucrats which has seen significant changes since 2001.

Westminster Comparative Perspective

To begin, the first school of thought entitled the Westminster Comparative perspective, perceived that the American Occupation following World War II had not only succeeded in eliminating the power of the military and family holding companies (*zaibatsu*), but had literally remade all of Japan's previous political traditions in Britain's image. The prime minister was believed to be a powerful executive, which was reinforced as a result of the Liberal Democratic Party's (LDP's) long reign as the majority party from 1955–1993. This comparison of Japan with Western states was taken to the extreme with scholars such as Arend Lijphart (1984, 38, 41) classifying Japan as "culturally homogenous" while asserting without empirical support that the Japanese government is "reasonably responsive to the citizens' wishes over a long period of time."

By the 1980s, Japan was thought to have been completely remolded into a Western system of government by a number of observers. For instance, Steven Reed (1993, 24) asserted that "Arend Lijphart has analyzed the industrial democracies of the world . . . and finds Japan to be quite similar to Italy." This was so even though "[t]he most important feature of Italian political culture has been its heterogeneity and fragmentation" (Hancock, et al. 1993, 307) whereas Lijphart was arguing the opposite. The point was that Reed (1993, 24) believed Lijphart had demonstrated Japan was a "normal country."

As the author noted nearly two decades ago, "[i]n their defense, these writers based this theory on two rather compelling pieces of evidence. . . . [T]he Japanese Constitution provides support for the formalistic interpretation that Japan had [in fact been turned into a typical parliamentary, western system of government]. . . . According to the Constitution, Japan is a parliamentary system with a bicameral legislature or Diet: an upper chamber called the House of Councillors and a lower chamber labeled the House of Representatives. Just like the British system where the upper chamber, House of Lords, [has been traditionally] subordinate to the lower chamber, House of Commons, the lower house in Japan has the final say on passage of legislative bills. This inability of the upper chamber to [ultimately] thwart the legislative desires of the lower chamber thus reveals that the house of Councillors 'is rarely able to play an independent role' (Ozawa 1994, 73).[2] Furthermore, the Constitution seems to create a fairly powerful prime minister's office. Though the prime minister is selected by the Diet as in England and is not directly elected by the people, '[t]he prime minister at the top of this structure is in theory all but omnipotent' (Ozawa 1994, 22). Once elected, the prime minister forms a cabinet with a majority of the cabinet ministers selected from the Diet. These cabinet ministers are assigned to head the ministries that have varied responsibilities such as the Ministry of Finance (*Zaimusho*) and the Ministry of [Economy], Trade, and Industry [METI]. When the prime minister resigns, all of the cabinet ministers must also resign" (Lebo 1999, 20–21).

Arend, Lijphart, and others did not grasp that the "1955 system" as the fifty-year, postwar tenure of the LDP came to be known, looked nothing like the above model. Ozawa Ichiro, formerly a high profile LDP Secretary General (Stockwin 1999, 73) who with Hata Tsutomu, broke from the LDP in 1993 to form the (now defunct) Japan Renewal Party (*Shinseito*), became a key DSP leader in the eventual overthrow of the LDP, and more recently broke away with the largest faction in tow from the DPJ in July of 2012 to form yet another party (BBC 2012), well describes this model. From 1955–1993, "[t]he prime minister [was] nothing more than master of ceremonies for the ritual at hand" (Ozawa 1994, 25). Rather, the Japanese prime minister faced a variety of constitutional, practical, and legal constraints on the exercise of formal authority. First, Article 4, Clause 1 of the constitution required collective decision-making by the prime minister and his cabinet with Article 6 further specifying that this decision-making be via unanimous consent (Takayasu 2005, 165–68). Although some disagreement existed among Japanese scholars over whether the prime minister might be able to make some decisions unilaterally due to the power to exercise control and supervision over administrative branches in Article 72 (see Takayasu 2005), in practice the prime minister faced significant legal constraints on the exercise of power (Shinoda 2011, 50–51).

In addition to having to share power and decision-making with cabinet officials, the Japanese prime minister's office was severely crippled by a lack of support staff and access to other sources of information most commonly available to prime ministers in other parliamentary systems. As Ozawa (1994, 47) clarifies, "[t]he staff [was] barely [able to] keep up with the routine tasks, let alone take the time to make future oriented plans for prime ministerial action." Additionally, the prime minister lacked even the power to select his own advisors (Ozawa 1994, 51). Indeed, the bureaucracy extended its influence into this arena for "the main source of advice to the prime minister's office on economic and financial matters is [a Ministry of Finance] official on temporary assignment" (Hartcher 1998, 258; Lebo 1999, 21–22).

The prime minister's cabinet fared little better as it struggled to exert influence over the prerogatives of the ministries. As Peter Hartcher (1998, 293) observed in the late 1990s while visiting the Ministry of Finance,

> The administrative vice ministers . . . meet the day before each scheduled Cabinet meeting. The bureaucrats set the agenda for the Cabinet meeting, debate the issues, make decision on each agenda item, and then draft the terms of the debate that Cabinet ministers will follow at their meeting. By the time the Cabinet meets, all decisions are made and agreed on at the bureaucratic level. The ministers act out their parts, reading out prepared notes and going through the motions of executive government, and endorse the decisions made the day before . . . [Therefore] Japan's Cabinet meetings are exceedingly brief, typically lasting some fifteen to twenty minutes.

As if these points were not enough to cause the casual observer to question whether Japan ought to be considered a typical, Western parliamentary system at all, one need only glance at the extreme turnover rate Japanese prime ministers have experienced to wonder at the efficacy of the prime minister's office. As J. A. A. Stockwin (1999, 97) explains, this means prime ministers had "little time to impose their personal imprint." As such, Ozawa Ichiro (1994, 48) concluded that "the prime minister is simply not in a position to exercise leadership."

While there is certainly a degree of truth in Ozawa's harsh indictment, it nonetheless fails to admit that in the absence of formal legal or constitutional authority, innovative prime ministers have found ways to exercise real power through informal means. Indeed, a growing literature indicates that as early as the 1940s when Prime Minister Shigeru Yoshida served as the first prime minister under General MacArthur during the Occupation of Japan, prime ministers have found ways to influence policy outcomes (Inoguchi 2011, 18–20). Shinoda (2011), for instance, provides a list of salient internal and external sources of informal powers that prime ministers have employed over the years to accomplish substantive policy goals.

Indeed, the powers exercised by various prime ministers have received significant attention in the literature. For instance, Inoguchi (2011) includes Shigeru Yoshida (1946–1947) and Tanaka Kakuei (1972–1974) as examples of effective prime ministers whereas Shinoda (2011) makes a similar argument focusing on the Yasuhiro Nakasone (1982–1987) and Ryutaro Hashimoto (1996–1998) administrations. Both studies conclude that prime ministers are capable of wielding significant influence due to such factors as their ability to exercise factional leadership within the party, ties with the opposition party, public support, business support, U.S. support, and capacity to elicit support from the bureaucracy. On this last point, Shinoda (2011, 52) notes that a string of the most successful prime ministers "boasted extensive personal connections with many bureaucrats in various agencies."

However, first amongst these case studies of prime ministerial power is the "presidentialized" (i.e., "one where the party leader becomes a more autonomous political force") tenure of Junichiro Koizumi (Krauss and Nyblade 2005). Koizumi enjoyed a number of sources of power, some of which were newly gained by the prime minister's office. As to the latter point, Koizumi was the first prime minister to be in a position to take advantage of the reforms of the earlier Hashimoto administration. As Shinoda (2011, 57) lists these changes which went into effect in 2001,

> Hashimoto's administrative reform brought major changes to enhance the political leadership of the prime and the cabinet. . . . First, the Cabinet Law was revised to bolster the prime minister's ability to initiate policy. Second, the functions of the Cabinet Secretariat, which is equivalent to . . . Britain's 10 Downing Street, were also reinforced with the authority to plan and draft important national policies. Third, the Cabinet Secretariat was enlarged, and the number of assistants to the prime minister was increased from three to five. Fourth, . . . [the law] placed the Cabinet Secretariat above other ministries and agencies, and strengthened the prime minister's power to coordinate policies among ministries. Fifth, the revised Cabinet Law identified fiscal and budget policies as important national issues, and authorized the prime minister to form the Council on Economic and Fiscal Policy (CEFP) under the Cabinet Office as an advisory organ independent of the bureaucracy to advise the prime minister on macro-economic and fiscal policy issues.

As this reveals, the prime minister's office in Japan enjoyed notably enhanced powers starting in 2001.

In addition to these impressive new formal powers, Koizumi also had an unusual array of informal powers to call upon which enhanced his power to influence the national policy agenda. First, he came to office through more unusual circumstances as he was elected by local party branches rather than through the traditional factions (Shinoda 2011, 57). This provided an independent source of support. Second, Koizumi also understood the importance of directly appealing to the electorate for support of his policies (Halloran 2007, 4). To do so, he carefully crafted messages that were easy to understand and support (Halloran 2007, 4). Other methods of exercising power included seeking alliances within rather than against the bureaucracy to help shepherd his preferred legislation through the Diet. As Isao Iijima, Koizumi's chief of staff and aide for thirty-five years explained in a 2008 interview, "Prime Minister Koizumi treasured the bureaucracy, and thought they had to be used properly" (Iijima 2008).

Likewise, Koizumi was not afraid of breaking with tradition even when it meant damaging his support within the LDP factions. For instance, he appointed his personal friend, Heizo Takenaka, an economics professor from Keio University, to serve as the Minister of State for Fiscal Policy among other posts, rather than relying on the classic selection of individuals to appease factions within the party. In addition, Koizumi became known for failing to consult with the Policy Affairs Research Council PARC (see infra) instead preferring to initiate his own policy proposals even if this meant "that he would pick a fight to see 'whether the LDP crashes the Koizumi Cabinet or the Koizumi Cabinet crashes the LDP'" (Shinoda 2011). As a result, in addition to his new formal powers, Koizumi liberally deployed "bypass strategies, namely initiating policies through prime ministerial advisory councils, building public support for those policies, and then on the basis of the reports and recommendations of those bodies, trying to bend the LDP and the bureaucracy to [his] will on a top-down basis . . ." (Mulgan 2003, 85). Although he met with mixed results (Mulgan 2011, 264), Koizumi's brinksmanship set a key precedent that the prime minister perhaps more today than ever before can marshal significant powers to sway policy in his preferred direction.

In fact, the current administration of Shinzo Abe (2012–present) appears to be proof positive that Koizumi was not an aberration. As the third longest serving prime minister at the time of this writing, Abe had proven to be a force to be reckoned with. His strong popularity led him to exercise substantial power over the bureaucracy. As one source notes, this is at least partly due to the long tenure of his key appointees including Yoshihide Suga, the longest serving Chief Cabinet secretary who works to coordinate the prime minister's office with the ministries along with the long tenures of Labor Minister Yasuhisa Shiozaki and Finance Minister taro Aso (Winter 2016). Thus, dismissing prime ministers completely is unwise.

Given these informal channels of influence open to prime ministers, this might lead one to inquire whether parliamentarians more generally are also capable of playing powerful policymaking roles? On the one hand, just like prime ministers, historically Diet members were also severely limited by resource and time constraints. Specifically, parliamentarians had little support staff and often relied upon other means of gathering information such as bureaucratic insiders or "moles" (Ramseyer and Rosenbluth 1993, 113–15).

These methods often left parliamentarians at a disadvantage resulting in receiving critical policy information too late to make a difference on the outcome (Ramseyer and Rosenbluth 1993, 115). However, on the other hand, just like prime ministers, writing off politicians completely would be equally unwise.

The main source of power for LDP politicians seeking to influence policy and bring home needed benefits for their home districts was through the party structure itself. The Policy Affairs Research Council (commonly known as "PARC")[3] served as a key intermediary between politicians and the bureaucracy. Prior to the approval of any policies by the Cabinet or being introduced on the floor of the Diet, all initiatives were required to be vetted through various substantive policy committees in PARC to thereby receive the LDP's endorsement (Fukumoto 2011). Over time, PARC provided a significant "medium for LDP Diet members to get the credit for delivering policy benefits to their supporters" and allowed politicians to gain much needed policy expertise (Mulgan 2003, 78). Likewise, the endorsement of the PARC chairman was the third most important for reelection of politicians after the faction of the prime minister and the secretary-general of the party (Fukumoto 2011, 55).[4]

The process became so formalized and regularized that a seniority system within PARC was initiated requiring that junior politicians work their way up through PARC's committee structure if they wished to ultimately achieve a high ranking political office such as a minister level appointment in the Cabinet or even to become a prime minister (Fukumoto 2011). In many ways, this system is foreign to observers of the Westminster system as it effectively allowed legislative back benchers to intervene in policymaking (Mulgan 2003, 80). As Aurelia George Mulgan (2003, 80) explains, "[t]he ruling party, not the executive, [was] the only political institution with sufficient power to bargain and negotiate with bureaucrats. Policy originates in the bureaucracy and is then subjected to political intervention by the PARC. In this process, bureaucrats and party politicians deal with each other on a direct and equal basis."

As such, the party itself served as an important counterweight to the exercise of bureaucratic power. At the same time, this system encouraged cozy relationships between bureaucrats, backbenchers, and business interests. Specifically, for a politician to gain in policy expertise eventually becoming a recognized member of a policy tribe (*zoku*) (Schoppa 1991), not only did the politician need to become engaged with committees in the PARC, but also needed to cultivate personal connections within the bureaucracy itself. As a result,

> [I]ndividual politicians . . . directly approach[ed] bureaucrats for favours and policy concessions in those areas that are within the purview of ministry offi-

> cials to grant. The bureaucrats can deliver the requested benefits because they have the power to do what the politicians ask without requiring ministerial approval. The strength and autonomy of the politicians is thus tied to the strength and autonomy of the bureaucracy. The result is a plethora of direct, lateral connections between LDP politicians and ministry officials which altogether bypass the [appointed] ministers in the prime minister's Cabinet (Mulgan 2003, 80–81).

This unique approach to governance is clearly antithetical to the classical Westminster model which envisioned a far more hierarchical policymaking structure with clearer chains of delegation.

Japan Inc. Perspective

Perhaps unsurprisingly, therefore, the Westminster school of thought soon found itself under assault as it was unable to explain Japan's phenomenal economic growth. As John King Fairbank (1989, 861) explains, "[b]etween the early 1950s and the early 1970s, the Japanese economy averaged an 11 percent yearly growth rate." Soon enough, a new school of thought arose to explain this phenomenon called the "Japan, Inc." perspective. Quite unlike the Westminster Comparative perspective, this model saw Japan as the polar opposite of a Western democracy. It argued that the only way these growth rates could have been achieved was through collusive relationships between the LDP, bureaucrats, and business dubbed "iron triangles." This first case of what came to be known as the "Asian Miracle" was thus taken as a sign that Japan was quite unlike a Western democracy. Moreover, Japan's alleged homogeneity was now seen as partly to blame. Indeed, "[t]he distinctive group-oriented behavior of the Japanese" was said to derive from "uniquely Japanese cultural attitudes" (Kohno 1997, 15).

This dusted off version of mercantilism proved, however, to be an equally poor fit to carte blanche describe the "Japanese" system. Rather, it more accurately described simply the long tenure of the LDP as the majority party in the "1955 system." Without going into full detail as to why this model should be discarded (see Lebo 1999, 30–35 for an explanation), a few words will help to clarify what led the proponents of this perspective astray. First, "Japan is not, nor has it ever been, a homogenous country on the Lijphart model. . . . Japan is a pluralistic society with caustic ideological cleavages that actually divide the society and hinder the economic intervention of Japan's constitutionally empowered bodies of government. To demonstrate Japan's heterogeneity, one need only observe the political party system where recognizable intra- and interparty ideological divisions are evidenced throughout the period of the LDP's dominance (1955–1993). . . .[With regard to] intraparty splits, for example, the Japanese polity can be best explained as one of 'divided politics' (Stockwin 1999, 4) due in part to the revival of vibrant multi-party competition in the 1960s. In addition to the LDP, the Japanese Socialist Party (JSP), and the Japan Communist Party (JCP), the 1960s saw the emergence of new parties including the Democratic Party of Japan (DPJ), the Clean Government Party (CGP), the New Liberal Club (NLC), and the Socialist Democratic Federation (SDF). While this reemergence of competing political interests was due to a number of factors, it should be understood to some degree as an expression of differing political perspectives. As Masaru Kohno (1997, 125) explains, "[t]o the extent that the two existing parties, the LDP and the JSP, were so far apart ideologically, there was enough space for the smaller parties to enter the middle." Moreover, while many have trumpeted the long reign of the LDP as a counterexample to the above observation, this argument also does not bear close scrutiny. Although the LDP retained a majority from 1955–1993, it hardly won through landslide victories. In fact, from 1972 through 1990, the LDP captured a high of only 58.98 percent in 1986 of the seats in the House of Representatives and a dangerous low of 50.49 percent in 1983 (Meyer and Naka 1998, 270). Based on the figures provided by Meyer and Naka (1998, 270), the average number of seats the LDP retained in the lower house was a mere 54.33 percent" (Lebo 1999, 30–31). This hardly suggests the Japanese citizenry walked in lockstep agreement on basic social issues. Rather, the Japan, Inc. model to the extent that it described the iron triangles between the LDP, bureaucrats, and businesses was really only observing the consequences of longstanding majority rule by a single party in Japan rather than a fundamental difference in the Japanese political psyche. These observations and many more misunderstandings wrought by the Japan, Inc. Perspective reveal that it ought to be discarded.

Bureaucracy Dominant Perspective

With the Westminster Comparative Perspective and Japan, Inc. Perspective thus retired, the next and probably most famous major explanation for Japan's success offered by Chalmers Johnson (1995, 123) was that the bureaucracy was the true power in Japan,

> In Japan, despite the fact that the constitution gives [the power to make laws] exclusively to the elected members of the Diet, it is the bureaucrats who actually initiate and draft virtually all important legislation. They also contribute significantly to the passage of bills within the Diet and possess extralegislative ordinance powers that are almost on . . . par with the statutes themselves.

In short, Johnson's (1982, 316) argument may be summed up neatly by a single line, "the politicians reign and the bureaucrats rule." Even more specifically, he argues that one ministry in particular—the former Ministry of International Trade and Industry (MITI)[5]—held the most power over Japan in the post–World War II Era. Without delving into Johnson's step-by-step analysis of how he believes the bureaucracy came to hold this power over the Japanese polity, the trouble with Johnson's rendition lies not so much in an error of commission, but in his errors of omission.

In addition to his failure to recognize the importance of PARC as a powerful partner in policy formulation along with the informal channels of power wielded by the prime minister discussed above, Johnson failed to recognize the divisions within the bureaucracy itself. As Stockwin (1999, 96) explains,

> It would be easy to conclude . . . that bureaucrats run politics, but in fact . . . the reality is a great deal more complicated than this. For one thing the persistence of interministerial rivalries brings about acute jurisdictional and policy disputes, so that the bureaucracy ought not to [be] regarded as a monolithic whole.

More details on these interministerial rivalries will become apparent in the context of the case studies found in chapters 2 through 6. In addition, the superficial nature of Johnson's argument became clearer in the subsequent decades as narratives of a single dominant force ruling the Japanese polity increasingly emerged. Whereas Hartcher (1998) insisted that the Ministry of Finance (MOF) was actually the key powerbroker in Japan's postwar economy, Ramseyer and Rosenbluth (1993) argued that the Diet was quietly controlling the bureaucrats. Whereas Schoppa (1991) focused on "tribes" (*zoku*) of legislators in the Diet who had formed powerful alliances around key policy issues, Gerlach (1992) detailed the powers of business agglomerations called *keiretsu* which had replaced the *zaibatsu* of the pre–World War II era. The point is simply that Johnson's search for the *one* most influential powerbroker that might explain how the Japanese polity operates, spawned a generation of similar efforts, which all reached different answers. This near mythical search for a single dominant power in Japan was misguided as the pluralistic nature of policymaking in Japan described above meant the absence of a single, dominant player.

EVOLUTION OF THE RELATIONSHIP BETWEEN BUREAUCRATS AND POLITICIANS

Given the multiplicity of theories for who controls Japan which raged from 1955 through the 1990s—even after the first turnover of power in 1993 of the LDP—the question becomes whether anything has changed since then? Is Johnson's bureaucracy dominant theory still in ascendancy or has another new paradigm begun to emerge? Does the relationship between bureaucrats and politicians described by Mulgan still attain or is there now a new status quo?

The answer is that changes have been underway since at least the 1990s which partially stem from significant reforms. First, as discussed above, the executive office in Japan has been significantly strengthened allowing the prime minister to exert more top down authority. Second, today political leadership and substantive policy expertise on the part of elected officials have become increasingly salient considerations at the ballot box. Third, classic strongholds of bureaucratic power have been dismantled or remade. The following, therefore, highlights the evolution of the relationship between bureaucrats and politicians in Japan since 2000.

The first sign that the relationship between bureaucrats and politicians might be in for change was the passage of electoral reform in 1994 when the LDP was temporarily out of power for the first time since World War II (Maeda 2009, 47). Importantly, by introducing a dual system of proportional representation (PR) and single member districts (SMDs), political hopefuls were suddenly competing in new ways,

> Under the new electoral system, the seats of the House of Representatives are divided into 2 groups: 300 members are elected from the same number of single-member districts (SMDs), and 200 (reduced to 180 beginning with the 2000 general election) are from 11 regional [proportional representation] PR blocs. Each voter casts a ballot for an individual candidate running in the SMD and one ballot for a political party that registers a list of candidates in the region. Candidates can run in both the SMD and PR tiers . . . those who secure seats in SMDs are automatically removed from the party lists for the PR seats. Candidates who are defeated in SMD races but win seats in PR are often called "zombie legislators" (Maeda 2009, 48).

As a result, significant prestige became attached to seats that politicians in SMDs won outright whereas zombie legislators were seen as consolation prize losers.

While initially voters under the new system seemed able to divorce their feelings regarding their SMD candidate from the party, by the mid-2000s, new trends began to emerge (Maeda 2009). First, a number of the smaller political parties folded leaving in place two major parties vying for dominance including the classic Liberal Democratic Party (LDP) and the newcomer Democratic Socialist Party of Japan (DPJ). Second, the two parties began to emphasize policy leadership at the party level more and more beginning with their reliance on "party manifestos" starting in 2003 (McElwain and Reed 2009, 289). As a result, the tension between candidate and party preferences began to dissipate (Maeda 2009) since politicians realized their vulnerability in the more competitive SMDs and concomitant need to evidence leadership and policy savvy at election time. The expectation, therefore, was that the new system would create significant pressure on politicians to think more clearly about national rather than just parochial interests.

A second key structural change that came six years after electoral reform was the passage of significant bureaucratic reforms. First, the bureaucracy experienced significant downsizing and Japan's ministries underwent massive reshuffling. Beginning in the year 2000, after recommendations from specifically designated task forces, the twenty-two ministries were reduced to eleven and experienced significant personnel cuts. In fact, the total number of regular service employees in the bureaucracy fell from 502,487 in FY1995 to just 280,316 in 2009 (see table 1.1 for all ministries). Rather than slowing down, additional reform proposals gained steam a decade later when the DPJ promised in the 2010 election campaign to continue reforming the bureaucracy. Some of these promises involved cutting personnel costs by 20 percent or 1.1 trillion yen, a process that has proved far more difficult to achieve than anticipated.

Table 1.1. Number of Regular Service Employees in the Japanese Government under the NPA Pay Scheme (2000–2009)*

FY	Total	National Personnel Authority	Ministry of Internal Affairs and Communications (Formerly: Ministry of Public Management 2001–2004)	Ministry of Foreign Affairs	Ministry of Finance	Ministry of Economy, Trade, and Industry	Ministry of Land, Infrastructure and Transport	Ministry of Health Labor and Welfare
*2000**	*496,315*	*683*	*6,455*	*5,207*	*14,658*	*8,167*	*49,028*	*80,837*
2001	484,099	708	6,089	5,152	14,681	*4,441	47,418	81,732
2002	481,905	699	6,030	5,191	14,870	4,687	46,234	81,598
2003	479,574	713	5,196	5,208	14,927	4,679	45,838	81,416
2004	302,839	714	5,187	5,243	15.038	4,645	45,405	38,011
2005	302,299	704	5,075	5,362	15,168	4,453	45,018	37,951
2006	300,034	707	5,009	5,334	15,113	4,392	44,334	37,564
2007	297,222	713	4,937	5,478	15,118	4,382	43,607	37,329
2008	292,405	719	4,899	5,489	15,210	4.343	42,877	37,609
2009	**280,316**	**730**	**5,114**	**5,590**	**15,431**	**4,343**	**42,510**	**38,377**

FY	Total	Ministry of Agriculture, Forestry, and Fisheries	Ministry of the Environment	Ministry of Justice	Ministry of Education	Ministry of Finance
*2000**	*496,315*	*21,486*	*1,121*	*46,773*	*135,127*	*14,658*
2001	484,099	15,751	968	46,820	136,101	14,681
2002	481,905	15,688	1,016	46,725	135,350	14,870
2003	479,574	24,369	1,058	46,758	134,933	14,927
2004	302,839	23,867	1,118	46,996	1,976	15.038
2005	302,299	23,328	1,146	47,128	1,975	15,168
2006	300,034	22,693	1,176	47,097	1,989	15,113
2007	297,222	21,372	1,193	47,351	1,987	15,118
2008	292,405	20,031	1,188	47,414	1,998	15,210
2009	**280,316**	**18,837**	**1,198**	**47,705**	**1,992**	**15,431**

Source: Courtesy of the International Affairs Division of the NPA.

*Japanese government reorganized ministries and agencies in January 2000. The figures only include regular service employees under the NPA pay scheme, excluding Special Service employees in the Ministry of Defense, the Diet, and the Courts.

(*Yomiuri Shimbun* 2011). Political actors are also revisiting the very nature of public service returning once again to consider whether public servants should continue to be denied the right to strike and collectively bargain in favor of vesting this authority in the NPA or any other body. This meshes well with the political platform of the DPJ which finds a receptive international audience in China when adopting the rhetoric of worker's rights (*International Herald Tribune* 2009b).

In addition to personnel cuts, the bureaucracy has experienced various challenges to its traditional sources of authority. For instance, the power of the bureaucracy to ensure that its employees received plum positions in the private sector upon retirement called *amakudari*, or "descent from heaven," has been in decline. Specifically, enhanced restrictions and oversight on the retirement options (*amakudari*) of bureaucrats (Reed and Shimizu 2009, 22; Maclachlan 2009, 161), along with restructuring of the workforce to include more temporary staff with limited tenure of employees under Koizumi's deregulatory policies (Nakamura 2012, 250), were largely to blame. All of these reforms and more are being attempted in the name of the "public good" to eliminate an alleged culture of bureaucratic siloing, "sectionalism," and "turf consciousness," which even rival political camps concur means "[t]he bureaucracy cannot function for the benefit of the public unless this mindset is changed" (*International Herald Asahi* 2009a).

While the continued power of the ministries cannot be denied, the mere fact that elected officials have been able to consistently wield such significant power to reform and restructure the bureaucracy over the past decade is instructive. The question becomes, however, what role the agency specifically tasked with implementing these reforms to curtail bureaucratic drift has played in this process? The next thus turns to evaluate the National Personnel Authority (NPA).

History of the National Personnel Authority

The history of the NPA bears review particularly in the context of events occurring in the United States in the aftermath of World War II. Although many studies have been undertaken of both of these topics, juxtaposing the comparative administrative history of the United States and Japan during this timeframe has not to the knowledge of this author been undertaken. Thus, the following describes the parallel events leading up to and transpiring during the creation of the NPA in both Japan and the United States.

To begin, the NPA was the invention of American designers seeking to ensure that the new democracy produced policies through a rational, objective decision-making process. In this way, it was hoped that the NPA would reduce societal competition while bolstering the merit based recruitment, training, and retention standards for government administrators. This is not to say that the concept of an expert, rational bureaucracy was somehow foreign to Japan or to Asia. Japan had experienced advanced bureaucratic administration extending at least as far back as the Tokugawa Shogunate (1600–1867 A.D.) while China invented the first civil service examination system during the Han Dynasty under Emperor Wu Ti as early as 136 B.C.E. (Fairbank, Reischauer, and Craig 1989, 169; see Schrecker 1991, 37–38 for placement in historical context). Rather, the key point is that the American occupiers sought to reframe the foundational logic and overarching structure of the bureaucracy along entirely new lines.

To better understand the eventual goals that the Americans settled upon, it bears note that the U.S. civil service system itself had been experiencing tremendous adjustment and upheaval since 1938. As the current U.S. Office of Personnel Management (2011, 3) explains, 1938 was a watershed moment in American history when the federal government sought to eliminate the corruption of the spoils system for federal employment in favor of a professional, competent civil service system. Two executive orders signed by President Franklin D. Roosevelt in June of 1938 expanded the powers and resources of the Civil Service Commission to effectively oversee the U.S. federal civil administrative system (U.S. OPM 2011, 3). This was soon buttressed in August of 1939, when Roosevelt also signed the Hatch Act into law which clarified, *inter alia*, that federal employees with a few exceptions were not permitted to engage in political activities (U.S. OPM 2011, 3–4). The Civil Service Commission under this initial iteration had significant authority to oversee the personnel selection processes along with discretionary review of the various agencies of the federal government.

The professionalization of the civil service continued in 1940 with the passage of the Ramspeck Act,

> This [A]ct was sponsored by Robert Ramspeck of Georgia, long an outstanding advocate of merit principles in the civil service and later (1951–1952) Chairman of the Civil Service Commission. The Ramspeck Act paved the way for an unprecedented extension of the merit system. It also . . . established efficiency-rating boards of review. . . . In effect, the act authorized the extension of the competitive service to more than 182,000 permanent positions—almost all the non-policy determining positions in the executive civil service (U.S. OPM 2011, 5–6).

However, with the entry of the United States into World War II in December of 1941, this centralized system was immediately placed under extraordinary stress to ensure the alacritous recruitment of 3.2 million civilian personnel by the end of the war in 1945. Displeased at the possibility of delays, the War Department attempted to undermine the power of the Civil Service Commission through a direct appeal to Congress to permit it to make its own recruitment decisions based upon a national emergency logic (U.S. OPM 2011, 7,

8). This was staved off through the actions of Arthur S. Flemming, a commissioner of the Civil Service Commission, through his own direct lobbying of Congress (U.S. OPM 2011, 7). He not only defended the Commissions' effectiveness, but also argued that national security would be undermined by the War Department's plan "and pointed out the confusion that would result if Government agencies were permitted to compete against each other in the tight labor market created by the emergency" (U.S. OPM 2011, 7).

Although he won the day and the War Department's cooperation, the exigencies of the time proved overwhelming for the Civil Service Commission, which eventually abandoned centralization of the U.S. federal administrative system. As soon as World War II ended, the Civil Service Commission was faced with the task of reducing the enormous wartime civilian workforce. Under President Harry S. Truman's directives, Civil Service Commission's practice of delegating authority to the various federal agencies with light oversight became the standard operating procedure. As today's OPM (2011, 11–12) explains,

> New civil service rules, issued under Executive Order 9830 of February 24, 1947, replaced the temporary regulations on May 1, 1947. An attempt was made to continue, in the new rules, the best practices developed during the war. The policy of delegating authority to agencies to act in individual personnel matters without prior Commission approval—such as individual promotions, reassignments, and transfers—was confirmed and extended. These actions, however, were subject to Commission standards, and to inspection and post audit by the Commission for compliance with standards.

This brief rendition thus brings us to the exact timeframe and administrative *milieu* in which General Douglas MacArthur as Supreme Commander for the Allied Powers (SCAP) was contemplating how to appropriately craft the civil service administration system in occupied Japan in a manner that would best serve the needs of the Japanese people and the new state. Likewise, this backdrop highlights that SCAP could not rely upon any tried and true American formula for constructing an effective civil service administration system in Japan.

To further complicate the task, the Japanese were themselves deadlocked over the contours of an effective public administration system. Just as the United States had been in 1938, the Japanese bureaucracy had become so entrenched, that to carry out significant reforms had become all but impossible. The Japanese had been engaged in abortive efforts to reform the bureaucracy since 1936 to eliminate problems including "sectionalism." As Koh explains in detail, the first movement for a new personnel agency actually derived from Prime Minister Konoe (June 1937–January 1939) who sought civil-service reform distributing proposals in 1938,

> . . . to establish a central personnel agency, increase the scope of discretionary appointment, revise the higher civil-service examination, initiate a program for retraining officials, and abolish security of tenure for officials. Among other things, these measures were aimed at undercutting sectionalism, enabling the government to recruit capable people from a broader base, combating *hoka banno* (the omnipotence of law graduates), and rejuvenating the bureaucracy (Koh 1989, 33).

Even so, none of these earlier reform efforts accomplished their desired outcomes largely due to bureaucratic resistance from the former Home Ministry (Koh 1989, 33). Thus, the arrival of the occupying American forces, themselves steeped in a foundational and transitional phase of civil service administrative adjustment and reform in the United States, clarifies that this moment marked a critical juncture in an ongoing saga of administrative development in Japan.

Yet, before we can fully understand that saga, one more foundational issue must be addressed. Although the Americans had cause to recognize the value of a competent administrative system to serve in both war and peace as a result of ongoing U.S. administrative developments, choosing the bureaucracy as the best ally to build a democratic Japan was not preordained. A word is required to explain, therefore, why the Americans settled upon the bureaucracy as the most logical partner in the democratization process of Japan. Certainly, other possible allies were available including the military, the corporate conglomerates (*zaibatsu*), and politicians.

The answer is due in large part to the political exigencies of the time. Preliminarily, from 1945–1946, the United States did not even view rebuilding Japan as a viable option perceiving that it was to be "forever a second rate nation" revealing just how deep anti-Japanese sentiments ran in America that perhaps Japan could not be trusted to rejoin the league of developed nations due to its pugnacious proclivities (Sheridan 1993, 124–26). Thus, as is widely known, American policies in the immediate postwar period were aimed at systematically eliminating the old guard seen as responsible for the most apparently irresponsible and irrational official act of all, *viz.*, the bombing of Pearl Harbor (Dower 1986).

In fact, the only explanation that seemed salient at the time for Japan's attack on Hawaii to patriotic Americans was that Japan's former military, political, and corporate leaders must have made these dangerous public policy choices based upon personal motivations such as power or glory as in the case of Japan's deposed military personnel or profit as in the case of Japan's hoary corporate conglomerates (*zaibatsu*).[6] This perspective equally tarnished the political leadership which was seen as fairly weak and tainted by the previous regime since it was viewed as incapable of resisting these extremist and seemingly irrational calls for war (Dower 1986).

Unsurprisingly, therefore, the military was literally disbanded with the entire nation constitutionally prohibited under Article 9 from engaging in aggressive warfare, two-hundred thousand individuals prohibited from holding public office (Dower 1999, 82), eighty-three large family holding companies (*zaibatsu*) first undercut with a capital levy (Fairbank 1989, 822) and then dismantled with "trust-busting legislation" in the first wave of reforms, and politicians considered "ultranationalistic and reactionary" purged from government and forbidden from running for office (Sheridan 1993, 124).

In this atmosphere, only one possible indigenous partner remained to help the occupiers construct the new democracy: the bureaucracy. As John Dower (1999, 27) explains,

> Because the victors had no linguistic or cultural entree to the losers' society, they had little choice but to govern "indirectly," through existing organs of government . . . For all practical purposes, General MacArthur's super-government relied on the Japanese bureaucracy to carry out its directives, creating in effect a two-tiered mandarinate. When the Americans departed, the native mandarins carried on, stronger than they had been even during the war.

In fact, General MacArthur through his lieutenants, such as First Lt. Milton J. Esman and Blaine Hoover of the Hoover Commission, which was under the auspices of the SCAP, began to note the theoretical linkage between democracy and bureaucracy while recognizing the practical necessity of doing so. For instance, in the deliberations leading up to the enactment of the Amended National Public Service Act (NPSA) in 1948 which established the NPA, the Esman Memorandum asserts,

> Of all the major bulwarks of feudal and totalitarian Japan, only the bureaucracy remains unimpaired. The bureaucracy will definitely outlast the [O]ccupation and will play a decisive role of m[o]lding the future of Japan. . . . Modern democratic government requires a democratic and efficient public service. Merely to reform obvious abuses . . . will not provide the minimum level of efficiency necessary to democratic administration now that the police are no longer available to perform the operating function of government. . . . [A] thoroughgoing democratization and modernization of the civil service [through] . . . relentless pressure from this Headquarters [is imperative] (Koh 1989, 38).[7]

Clearly choosing to anchor the new democracy with Japan's extant corps of expert bureaucratic administrators was not without risks and may strike a discordant note to the modern inner ear particularly given Esman's admission that the bureaucracy had been the bulwark of a totalitarian state.

Even so, both recent experiences with civil service administration in the United States along with practical necessity won the day and in marked contrast to the other remnants of the former regime, MacArthur, the "blue-eyed shogun," ensured that only "840 of the 42,251 bureaucrats (1.9 percent) . . . were purged" including only 145 members of the senior civil service (Koh 1989, 38–39). Yet, while the bureaucracy would prove to be a powerful ally, the process of democratization and state building hardly proceeded in a lockstep or even in an uncontested fashion. Rather, MacArthur soon found that the ally that he had called upon was more recalcitrant than he had imagined.

Due to widespread economic dislocations in the fledgling democracy from a disastrous war, exacerbated in no small part by the dismantling of Japan's corporate conglomerates and the early policy of keeping Japan a "second-rate" nation-state, social unrest rose. With the passage of the Labor Union Act in 1948, national public employees, excluding certain groups such as police and fire service personnel, were suddenly guaranteed various rights including the permission to enter into agreements and bargain collectively along with the right to strike (NPA 2010c, 58). This culminated in significant social upheaval in the "July Struggle" of 1948 (NPA 2010c, 59), which is visually recorded in a number of documentaries as a significant challenge to the new nation with shockingly violent clashes between strikers and police. These circumstances promised to completely paralyze various key industries thus threatening the nascent new economy and, thereby, unraveling Japan's tentative steps on the path towards democracy. Even more disconcerting from the American perspective at the time was the observation that various *government* workers were also primed to exercise their newly granted labor rights and take to the streets thus shutting down the state and threatening domestic security. Importantly, it was during this period that communism was also on the rise throughout Asia as China was in the midst of civil war between the Chinese Communist Party under Mao Zedong and the KMT (Kuomintang) under Chiang Kai-shek. Likewise, trouble was on the horizon in Korea for what would soon escalate into the Korean War.

Hence, with the rise of socialist and communist parties in Japan, the United States became deeply anxious to ensure that democracy gained a permanent foothold in Japan. Therefore, anti-Japanese sentiments gradually changed from 1947–1950 as the United States became preoccupied with these global events concluding that Japan might serve as an effective shield against communism in the Pacific Rim (Sheridan 1993, 124–26; Borthwick 2007, 215). Again, MacArthur's SCAP perceived that this would only be possible in the long-run if the bureaucracy's mission and the stability of the new democratic state itself could somehow be inextricably fused.

As tensions worsened and the Government and Central Labor Relations Committee failed to achieve an agreement, MacArthur personally intervened in August of 1948 and

> ordered an amendment to the National Public Service Act to stress the peculiarities of public employees and deny the right to bargain collectively and the right to strike, and demanded that a temporary civil service commission serve as an organization responsible for protecting public employees' interests as a stopgap measure until the amendment was made (NPA 2010c, 58).

The outcome just four months later in December of 1948 was the much more permanent NPSA establishing the NPA and with it, Japan's modern public personnel system (NPA 2010c, 58).

While one might suppose that establishing the NPA therefore proceeded smoothly due to MacArthur's preeminent stature and power in Occupied Japan, the reality is that the NPA's creation and significant powers to represent the interests of the people and collectively bargain on behalf of the bureaucracy was anything but a foregone conclusion. Winding the clock back one year earlier to July of 1947 in the heart of administrative reform in the United States, Japan's Diet members had already begun airing their concerns over SCAP's plan for a national personnel agency. These politicians took great pains to underline the alleged ambiguities in the draft of the law to Blaine Hoover who was MacArthur's key lead on the project and acting chief of the civil service division (Koh 1989, 47, 50). In fact, the Diet members' substantive objections were almost as numerous as their stylistic ones. Hoover was seen as an unwelcome and unpleasant middle-man determined to effect changes without regard for Japanese sensibilities

Thus, the plan to create a single entity responsible for overseeing the public civil service system was met with carefully veiled defiance and resistance with the involved officials indicating at least ten key areas of perceived ambiguity in the intended authorizing legislation (Koh 1989, 46). The objections were difficult to ignore as they had been drafted by the ministries at the request of the Diet and MacArthur as SCAP had obligated himself to make leadership *recommendations* rather than issue edicts or decrees for public relations purposes (Koh 1989, 47).

As fate would have it, during this critical juncture when the draft of the law was being drawn up, Hoover took a leave of absence returning to the United States. As Koh (1989, 47–48) explains,

> The Japanese, recalling the maxim "*oni no inai aida no sentaku*" (doing laundry while the devil is away), worked feverishly to produce their own draft legislation incorporating all desired changes, both stylistic and substantive. In a covering letter accompanying the new draft sent to the government section on 31 July, [1947] Prime Minister Katayama minimized the importance of proposed changes, saying that they were either stylistic or minor. "Although [the revised draft] may appear to be considerably different from your original draft," he said, "it is not substantially different from [your draft] in terms of basic principles."

The key changes cited in his letter, nonetheless, were revealing. The power of the NPA would be reduced by, *inter alia*, omitting the provisions dealing with its finances that would have given it a measure of independence in budgetary matters; the NPA's power to enact regulations having the effect of law would be reduced; and key provisions dealing with the prohibition of strikes by public servants would be deleted (Koh, 1989, 47–48). When Hoover returned in November, he discovered that the NPA's power to regulate the bureaucratic system had been almost completely eviscerated by the newly enacted public service law a month earlier. Irate at this outcome, Hoover then resorted to more secretive tactics and communications with MacArthur establishing his own legislative drafting body to revise the law and restore power to the NPA as initially envisioned.

In the meantime, the severity of the economic and political crisis in Japan had steadily worsened through the middle of 1948, lending enhanced urgency to clamping down on possible government shutdowns due to labor strikes. In the end, MacArthur finally personally intervened on July 22, 1948. In contravention of his typically more diplomatic approach, to put a stop to the political battle between Hoover and the Diet, MacArthur ordered Prime Minister Ashida Hitoshi to comply with his "recommendations" (Koh 1989, 52). Ashida did so with great alacrity and thus established the NPA's authority in the form originally intended by the SCAP. The critical powers of the NPA were not only restored to Hoover's initial expectations, but expanded by the bill passed in December. As Koh (1989, 54) summarizes,

> Its rule making powers were enhanced: it now monopolized the power to make rules aimed at implementing the standards set in the national public-service law; the erstwhile requirement that all rules must be approved by the prime minister was jettisoned. The [NPA's] financial position was bolstered by the acquisition of the privilege of 'double budget': should the cabinet decide to revise the [A]uthority's proposed expenditures, it must submit to the Diet both the revision and the original request. [Moreover,] measures to tighten discipline in the public service were introduced. This meant, primarily, that public servants were stripped of the right to strike and to employ delaying and other dispute tactics, were forbidden to run for elective offices or become officers in a political party or any other political organization, and were barred from all political activities other than voting. As a result, the gap between public servants and private-sector employees widened considerably.

Thus, the blue-eyed shogun with a single letter established Japan's premier personnel agency tasked with overseeing the entirety of the administrative system. To emphasize this point, the NPSA passed in December 1948 created through *executive* rather than *legislative* fiat an institution designed to oversee a bureaucracy and ensure its fit within a parliamentary system. A more contested and paradoxical birth for an institution is hard to imagine.[8]

Yet, while the casual observer might be tempted to assume that executive fiat alone was sufficient to alter the course of Japan's administrative system, subsequent intervening global events also played a critical part. The NPA proved to be a quiet beneficiary of the dramatic turnaround of the Japanese economy due to the Korean War economic boom (Dower 1999, 542) and later the Vietnam War economic boom (Havens 1987, 92–97). As Tomas R. H. Havens (1987, 93) explains,

> United States procurement for the Korean War had hoisted a flat Japanese economy toward unexpected prosperity. In 1952, the peak year, 63 percent of all Japanese exports were taken by the war. Even after the cease-fire in 1953, payments by the United States in connection with its bases and personnel in Japan averaged nearly $600 million for the rest of the decade, and they still accounted for about 14 percent of Japan's exports as late as 1958–1959.

Thereafter, strong procurement of approximately $210 million in 1964 and annual U.S. expenditures continued to bolster the Japanese peacetime economy (Havens 1987, 93). However, with the Vietnam War, U.S. military procurement rose again to $292 million in 1967 and $467 million in 1970 accounting for between 7–8 percent of Japan's annual exports buoying the economy that had finally begun to flag in 1965 (Havens 1987, 94). As Murakami Yoshio, a cabinet member in the 1980s is quoted as saying, "the Vietnam War to Japanese conservatives was an opportunity, not a problem" (Havens 1987, 97).

Thus, as worker unrest subsided and the Japanese economy settled in for a long period of unprecedented growth, the NPA's existence was no longer a salient issue and the Authority quietly became a permanent and seemingly necessary fixture of the administrative state. Thus, even though MacArthur as SCAP had long since departed, the impetus to significantly change the foreign established personnel system for administering the state as the guarantor of democracy evaporated and the NPA was left with its expansive mission and powers largely intact.

Indeed, with the tradition and parallel histories of civil service administration in both the United States and Japan thus juxtaposed, the ultimate goals of the American occupiers now become clearer. First, the Americans sought to redefine the role of bureaucrats as "servants of the whole people" rather than the emperor. This new concept was soon enshrined in Article 15 of Japan's Constitution (NPA 2010, 56). Second, although the system had already been abandoned as unworkable in the United States, SCAP sought a single, powerful personnel agency responsible for setting the rules of the system and managing the civil servants to ensure that they properly understood their duties to serve the people. Thus, when the American occupiers left, the NPA remained as a monument to the occupier's vision of a new Japanese administrative state. The next section considers how the NPA has come to operate since then comparing its powers on paper to its powers in fact.

POWERS OF THE NATIONAL PERSONNEL AUTHORITY

To grasp the breathtaking intended sweep of the NPA's authorizing legislation, which is remarkably reminiscent of the U.S. Civil Service Commission noted above, it is instructive to consider how the NPA summarized its responsibilities sixty years later. As the NPA's 2010 Annual Report clarifies, the NPSA tasked the NPA with "ensuring fairness in the personnel management of public employees" while acting in the best interests of the public employees themselves by ascertaining their labor rights and compensation (NPA 2010c, x). Likewise, the NPA insisted it was explicitly founded "as the organization specializing in personnel administration [to] advanc[e] appropriate policies and measures in response to the situation in society in general aiming to ensure efficient public administrative operations that can be trusted by the public" (NPA 2010c, x). The expansiveness of this mission statement should capture the attention of even a cursory read. In particular, one must note that this mission statement proclaims a dualistic function including *both* the classic public administration's emphasis on efficiency *as well as* the political control theory's emphasis on responsiveness. As the discussion further on will reveal, however, in practice the NPA focused largely on efficiency and much less on responsiveness. At the same time and somewhat paradoxically, although the NPA continued to proclaim this broad mission statement right up until the 2014 reforms, the institution itself had often been dismissed in the administrative literature and anecdotally. Numerous scholars insisted that the agency was a weak shadow of its potential due to numerous constraints such as understaffing, borrowed staff, and restricted oversight powers (Colignon and Usui, 2003, 22–23, 47, 49, 178; Suzuki 2006b, 202).

While the capacity of the NPA to actually and effectively engage in systemic level activities will be considered shortly, the point remains that its authorizing legislation had not changed. For nearly 65 years after its establishment, the NPA remained the only administrative institution in Japan with systemic level responsibilities to define, manage, incentivize, and penalize "public servants" while overseeing the administrative personnel system. Only the NPA maintained communications and links with all eleven ministries, though it was not responsible for the Ministry of Defense, and reported directly to the prime minister. Although its training programs to enhance the responsiveness of administrative agents to the signals of political principals were continually scaled back in the years leading up to the 2014 reform, when it comes to all personnel issues, the NPA remained supreme. Thus, dismissing such an entity offhandedly is unwise.

Turning to the former powers of the NPA requires assessing how it actually operates today. In short, the professional life of the Japanese administrator started and finished with the NPA. Although not a ministry itself, the NPA was charged with administering the public service examinations, enhancing and evaluating bureaucratic efficiency, performance, and management throughout the ministries, promulgating and enforcing reporting systems and ethical standards of job performance to ensure transparency, building close ties with representatives of foreign governments through personnel exchange and training programs, and finally reviewing the retirement plans and joint appointment proposals of top level bureaucrats to ensure their legality (NPA 2010c; NPA 2006; NPA 2002).

The NPA thus served key coordinating functions for the entire system with a strong mandate to enhance the efficiency of Japan's system of public administration. For instance, the NPA acted as an important implementer of a number of recent reforms including cutting the total number of civil service personnel 25 percent by 2010, ". . . the introduction of self-evaluation policies[,] and the legislation of a civil service ethics law in 1999" (Suleiman 2003, 167). It was also responsible for overseeing the famous "*amakudari*" process whereby retiring bureaucrats may "descend from heaven" to take much better paying private sector jobs (Colignon and Usui 2003).

ROLES OF THE NATIONAL PERSONNEL AUTHORITY

While the NPA thus had important powers in the Japanese system, the question remains whether it was in fact carrying out the mission that MacArthur's SCAP envisioned for it all those years ago? In other words, by focusing on efficiency, was the NPA the guarantor of democracy that it was intended to be? Did the NPA in fact shape and constrain the discretion of administrators? Did the NPA achieve its constitutional mandate of ensuring that civil servants represent the whole people rather than the emperor? Or was it time for curtailing its powers and handing off some of these responsibilities to a new entity?

To evaluate the NPA's role, it bears recognition that this body is the legislature's official tool of *direct control* over the bureaucracy (Meier and O'Toole 2006). As such, the NPA as an agent itself used approaches that are deeply steeped in the hierarchies of efficiency school of thought. In the parlance of the public administration literature, the NPA came to envision its larger systemic role increasingly as the duty to train "responsible" meaning *efficient* administrators rather than to constrain bureaucratic drift. In fact, the literature of the NPA is almost a study in classic public administration efficiency thinking. The key point is that the NPA in action was the perfect exemplar of why political control theorists assert that direct control of political principals over administrative is difficult to accomplish and is generally undesirable to attempt. In other words, although designing a system of control is feasible, if predicated on efficiency, it is unlikely to be effective.

Turning to the public administration literature reveals that "responsibility" has four concrete components: (1) clearly defining and delegating duties, (2) providing resources to ensure the completion of those duties, (3) measuring and evaluating the performance of said duties, and (4) sanctioning the failure to comply with those duties within the specified parameters (Mosher 1982, 9). Rephrased, this means ensuring public servants are appropriately and thoroughly trained, resourced, evaluated, and managed. Japan codified these administrative responsibilities in the Administrative Procedure Act of 1993 (ch.1, art.2, §vi) as "guidance, recommendations, advice, or other acts by which an Administrative Organ may seek, within the scope of its duties or affairs under its jurisdiction, certain action or inaction on the part of specified persons in order to realize administrative aims, where such acts are not Dispositions." In short, in tune with the bureaucracy's nearly obsessive attention to detail, even the undefined discretionary duties have been defined (Cabinet Secretariat 1993, 2).

The modern roles of the NPA officially encompassed all four of these components of responsibility. The following considers the first two components of responsibility. Despite the significance of these responsibilities, due to budgetary constraints, limited staff, and the multiplicity of duties under its official powers above, the NPA was indeed unable to effectively accomplish this larger systemic mission. To evaluate the first (i.e., clearly defining and delegating duties) and second (i.e., providing resources to complete those duties) components of responsibility, the focus here is on the democracy building activities and personnel training programs instituted until 2014 by the NPA.

Component 1: Clearly Defining and Delegating Duties—The NPA's Democracy Building Activities

The NPA attempted a variety of foundational democracy building activities.[9] The NPA undertook these activities to gauge public sentiment towards administrators in Japan and thus refine and improve the NPA's promise to work towards achieving the public's expectations that administrators must act as "servants of the whole people." The findings of these projects and activities were used to justify continually revisiting and revising the training programs that the NPA directs for administrators throughout the system (see next section). To demonstrate this point, this subsection considers both the NPA's "2009 Monitor of National Civil Servants: Survey Results" (June 2009), which serves as a yardstick of public opinion towards public servants and the sundry activities in which the NPA is engaged, along with an NPA commissioned study by Takashi Nishio et al. (2010) of how to translate those findings into an affirmative action plan that effectuates the first component of responsibility: namely, duty definition and delegation for public servants.

To present how the NPA carried out this approach, the assessment undertaken by Professor Takashi Nishio, et al., in February 2010, was secured from the NPA and translated into English. The project was specifically commissioned to realign the mindset of Japanese administrators with the expectations of the general public. The title the project is telling, "Cultivating Professional Public Employees in the New Age—Image of the 'Servants of the Whole People' that Support the Political Leadership" (Nishio 2010). The Study Group on Public Service Training and Human Resource Development was carried out by the Research Society on the Development of Human Resources in coordination with the National Personnel Authority.

To construct reliable recommendations for changes to the implementation of public service training seminars to ensure that administrative behavior aligns with the public's expectations, the Nishio Study Group relied upon data such as a two week survey (i.e., 6/15/2009–6/24/2009) of a diverse cross-section of the population including 500 citizens over 20 years of age who were designated as "Monitors of National Civil Servants" and then questioned on various aspects of their expectations of public administration conducted by the National Personnel Authority (see table 1.1). For instance, one question asked, "Which areas should the NPA focus on to manage public service personnel objectively and fairly? (From five choices, select all that apply.)" The top three areas of concern for the respondents included (1) "planning and implementing the examinations for public service," (2) "observing political actions, separating political activity from private enterprise," and (3) "training and development efforts, planning, and implementation."

The results of this analysis were then carefully evaluated and recommendations for changes to approaches in the personnel management system were constructed. For instance, with regard to the findings of the study on the expectations for enhanced training and instructing public servants on the nature of their civic duties, the Nishio Study Group (2010, 9) discovered a "general deficiency in understanding what it means to be a public servant for the Japanese people." To "resolve" this issue, the Nishio Study Group recommended creating a "philosophical change . . . among all public service personnel, particularly among the top-level personnel" while the NPA "[c]ontinue[s] to reinforce the point after following initial employment that public servants work for the people" (Nishio 2010, 9).

Component 2: Providing Administrators with the Resources to Complete those Duties through NPA Personnel Training

The second key component of responsibility is providing management training to equip public servants with the organizational tools required to complete their duties. Taking the above observations and translating them into an affirmative personnel training program, while a daunting task, is one that the NPA attempted every year. To ascertain precisely how this transpired became apparent when examining training schedules and other documentation from the NPA which recorded how the recommendations from the Nishio Study Group and other yearly evaluations derived from the National Monitor survey data and other sources were carried out.[10] In addition, the experiences of the Mansfield Fellowship alumni once again provided a rich source of data. All eight of the selected participants in the institutional case study analyses in chapters 2 through 6 also completed the NPA personnel training courses. Likewise, additional Mansfield Fellows who participated in this project as research subjects also attended the NPA's personnel training and their observations are included as well to supplement and enhance the analysis as necessary. Importantly, the training program in place in FY2009–2010 is the model of the system assessed although this program underwent changes from 2000–2010 and continued to undergo shifts which were driven at first by budgetary considerations (Watanabe 2011, personal communication) and then by organizational changes wrought by the 2014 reform. Nonetheless, the model evaluated is a reasonable approximation of the general methods and approaches that were in place over this time period to train administrators.

To begin, the 2009 Monitor which evaluated the opinions of 500 citizens posed questions of residents from all four major islands including Hokkaido, Tohoku, Kanto, Chubu, Kinki, Chugoku, Shikoku, and Kyushu/Okinawa to ascertain their perspectives on topics ranging from the purpose of the NPA to their expectations of civil servants (see table 1.2). As these data are collected on an annual basis, they provide not only a guidepost for agency evaluation, but also a means for tracking the progress of the NPA in achieving those goals from a more longitudinal standpoint.

Turning to the questions posed by the survey, the familiarity of the respondents with the NPA itself was first ascertained (see figure 1.1). Of those surveyed, a mere 5.7 percent of the respondents had never heard of NPA while 59.9 percent were either somewhat or very familiar with the functions of the Authority. This is in itself remarkable as it demonstrates that the NPA had a high level of salience amongst the general public even though it was not engaged in a familiar policy context such as healthcare or trade policy. While neither 2009 Monitor nor the Nishio et al. Study Group say so, this provides prima facie evidence that the NPA was taken seriously by the gen-

Table 1.2. Demographic Characteristics (Monitors of National Civil Servants). Study Period—June 15, 2009 to June 24, 2009.

Gender	n=250 (50.0%)	Area	n=25 (5.0%)
Male	n=250 (50.0%)	Hokkaido	n=40 (8.0%)
Female		Tohoku	n=180 (36.0%)
Age Group	n=75 (15.0%)	Kanto	n=70 (14.0%)
20-29	n=100 (20.0%)	Chubu	n=80 (16.0%)
30-39	n=100 (20.0%)	Kinki	n=30 (6.0%)
40-49	n=100 (20.0%)	Chugoku	n=15 (3.0%)
50-59	n=125 (25.0%)	Shikoku	n=60 (12.0%)
60 and up		Kyushu/ Okinawa	
Occupation		n=45 (9.0%)	
Self-employed		n=152 (30.4%)	
Company employee		n=51 (10.2%)	
Part-time employee		n=34 (6.8%)	
Student		n=119 (23.8%)	
Housewife		n=66 (13.2%)	
Unemployed		n=33 (6.6%)	
Other			
Note: Other category includes doctor, nurse, farming industry workers, etc.			

Source: National Personnel Authority, June 2009, "2009 Monitor of National Civil Servants: Survey Results" (unofficial English translation by Professor Fredrick Butcher).

eral Japanese citizen if not by scholars on the topic (Colignon and Usui 2003, 22–23, 47, 49, 178; Suzuki 2006b, 202). If a government institution maintains a sufficient presence in the public arena of an advanced democratic state to merit the attention of the general citizenry, its activities at least warrant closer scrutiny.[11]

Upon securing this finding, the 2009 Monitor apprised the respondents of seventeen activities that the NPA engaged in and solicited their feedback as to which activities were of the greatest interest. Of the seventeen topics, the top three in order of interest included salaries, personnel evaluation, and recruitment procedures rather than the development of personnel (see figure 1.2). While not particularly surprising, these findings confirm that the focus of the general citizenry tends to be on the specific powers of the NPA rather than on its larger mission. Specifically, "development of personnel" categories which is the training process whereby standards of principled decision-making are relayed to administrative agents garnered between 27.1 percent and 28.8 percent of the interest of respondents. This ranked fifth being considered even less interesting than learning basic facts about administrators such as the numbers and types of civil servants. As such, when the respondents were asked to comment on the activities that the NPA should "aggres-

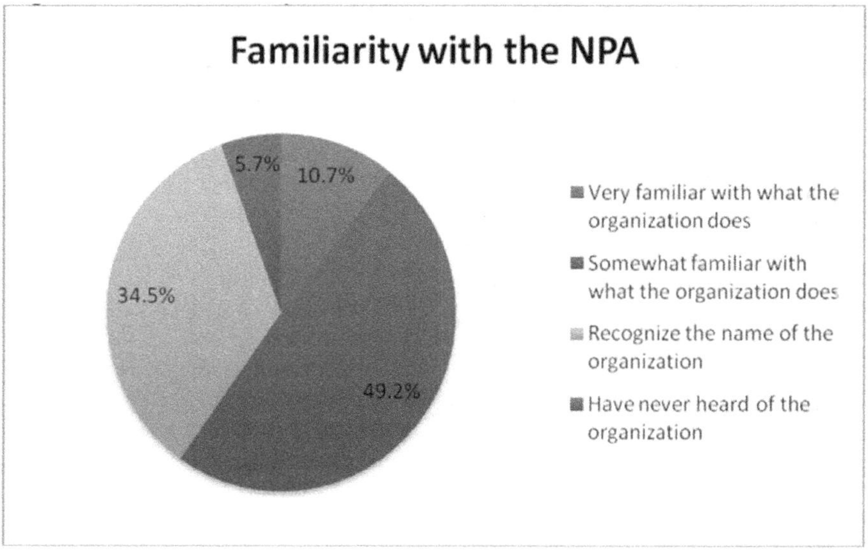

Figure 1.1. Familiarity with the NPA. Created by the author with reference to the National Personnel Authority of Japan. Unofficial translation by Professor Fredrick Butcher. Note: "Respondents' level of familiarity with the NPA prior to completing the questionnaire."

sively" focus on (see figure 1.3), the same specific powers of the NPA gathered more interest than its larger role in the Japanese democracy. For instance, the NPA's responsibility to build skills amongst administrators relevant to their jobs as civil servant came in fourth.

The next set of inquiries attempted to gauge the interest of the public more directly towards administrators rather than the responsibilities of the NPA. Comparing the 2008 Monitor to the 2009 Monitor, we find an overall increase in those who perceive that overall administrators either have a weak sense of ethics or are all unethical from 25.8 percent to 29.6 percent or a rise of 3.8 percent. Yet, these data grow far more illustrative when juxtaposed to the NPA's findings on the same question focusing on the subset of respondents who were employed at the time of the survey. The employed respondents generally provided more positive opinions of the bureaucracy than unemployed citizens (see figures 1.4 and 1.5). In fact, the percentage of respondents perceiving public employees are "overall" weak on ethics (meaning generally as a group) or that all public employees are weak on ethics shrank from 16 percent in 2008 to just 3.7 percent in 2009.

This suggests that the economic circumstances of respondents are linked to attitudes towards the bureaucracy. In other words, if the respondents were

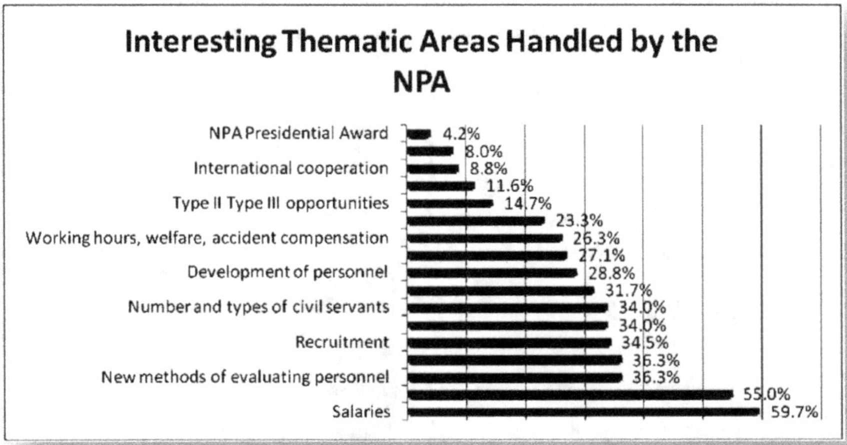

Figure 1.2. Interesting Thematic Areas Handled by the NPA. Created by the author with reference to the National Personnel Authority of Japan. Unofficial translation by Professor Fredrick Butcher. Question: "Among the many areas in which the NPA deals with the administration of public service personnel, which areas or themes would you be interested in finding more about? (From a total of seventeen choices, select five.)"

gainfully employed over the 2008–2009 biennium, they were increasingly likely to have a favorable impression of the bureaucracy with the reverse also being true. To some extent, therefore, the economic circumstances of respondents were linked to public perceptions of the bureaucracy with those who were jobless apparently more likely to find fault with the bureaucracy. This lends support to the assertion in the Introduction that the bureaucracy is either the scapegoat or beneficiary of the vagaries of the Japanese marketplace.

Considering the above results along with various other survey data not included herein, the Nishio et al. (2010) group concluded that while Component 1 (i.e., specialized training for civil servants in the area of expertise necessary to complete their duties as the respective agencies) was in order, Component 2 (i.e., "[g]eneral education and training for all public servants across the country") was not. This may at first seem curious considering that the findings above did not indicate dissatisfaction with or even much interest in the larger systemic role of the NPA. However, given that the purpose of the 2009 Monitor is ascertain the public's wishes for their civil servants and to craft personnel policies accordingly, the Study Group's emphasis on Component 2 (i.e., general education and training) is understandable.

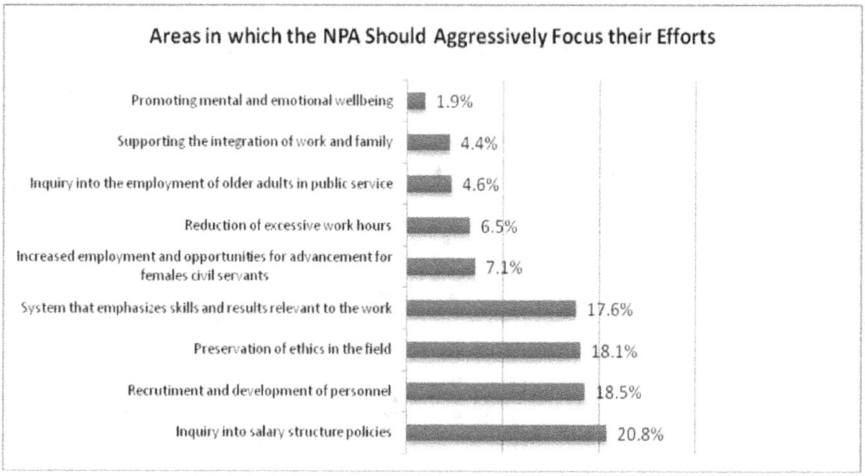

Figure 1.3. Areas in which the NPA Should Aggressively Focus their Efforts. Created by the author with reference to the National Personnel Authority of Japan. Unofficial translation by Professor Fredrick Butcher. Question: "Regarding the public service system, what is the one area in which the NPA should put the most effort into addressing? (Choose one of the nine choices.)"

The next step was to determine whether to tackle Component 2 by making recommendations. One clear lesson from the data above is that respondents were concerned that administrators adhere to ethical standards. Given the need to reassure the public after various scandals have rocked the ministries in previous years, Nishio et al. (2010) might have focused on recommending improvements to the general ethics training public servants receive through the NPA which are part of its formal powers.

Yet, the record of the NPA on this front is strong as are its approaches. For instance, the Ethics Act and the Ethics Code have been clarified for public servants since 2002 through an annual National Public Service Ethics Campaign Week in December which is replete with campaign posters, pamphlets, and slogans selected from those submitted by the public along with a Public Service Ethics Hotline set up in June of FY2006 to receive complaints not to mention a significant series of lectures at various locations hosted by experts and academics (NPA 2010c, 179–80).[12]

Thus, rather than recommending additional procedural controls or refinements to existing initiatives such as enhancements to ethics training for civil servants, Nishio et al. (2009) emphasized the larger systemic duties of the NPA concluding that "training and education [must] occur with a focus on fairness and long-term thinking." Again, this is in tune with the public administration's insistence that enhancing administrative efficiency requires

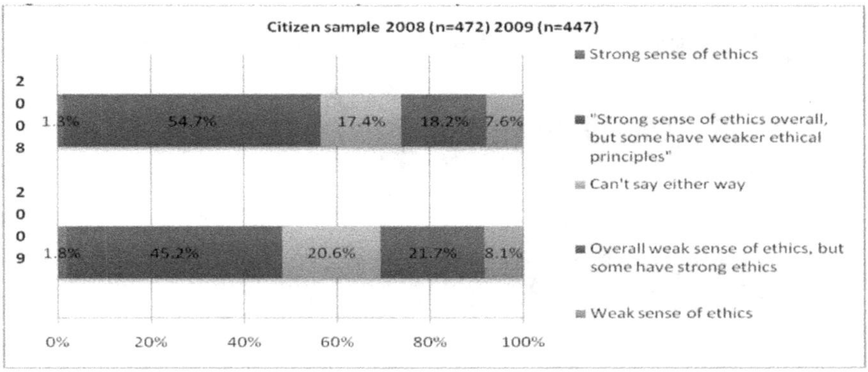

Figure 1.4. Citizens' Evaluation of Ethics of Public Administrators. Created by the author with reference to the National Personnel Authority of Japan. Unofficial translation by Professor Fredrick Butcher. Question: "What are your current impressions regarding the ethical principles of civil servants?"

focusing on the behavior of administrators. Nishio et al. (2010, 3) recommended that the NPA should pursue six changes to its training regimen including:

1. "Devotion towards becoming the ideal civil servant."
2. "Cultivating a sense of pride in the vocation."
3. "Maintain[ing] a separation from the political process."
4. "Cultivat[ing] a philosophy that parts from the traditional organizational structure to solve real world problems."
5. "Develop[ing] problem-solving skills that involve the analysis of the administration including administrative failures."
6. "A personality that matches with a classical form of education-ethics."

The fourth through sixth goals are particularly startling considering that the fourth aims to change the nature of decision-making itself in the bureaucracy to be more solution driven rather than institution bound. In like manner the fifth and sixth goals are focused on training public servants to employ enhanced problem-solving skills and to rethink their "personality" (Nishio et al. 2010). As this discussion reveals, at the NPA's request, Nishio et al. (2010) sought recommendations to strengthen the NPA's larger role in the Japanese democracy. Thus, the question becomes how these theoretical recommendations might be translated into concrete training programs which also tracks the second component of responsibility.

To evaluate the second component of responsibility—providing the requisite resources to administrators to complete their duties—the following con-

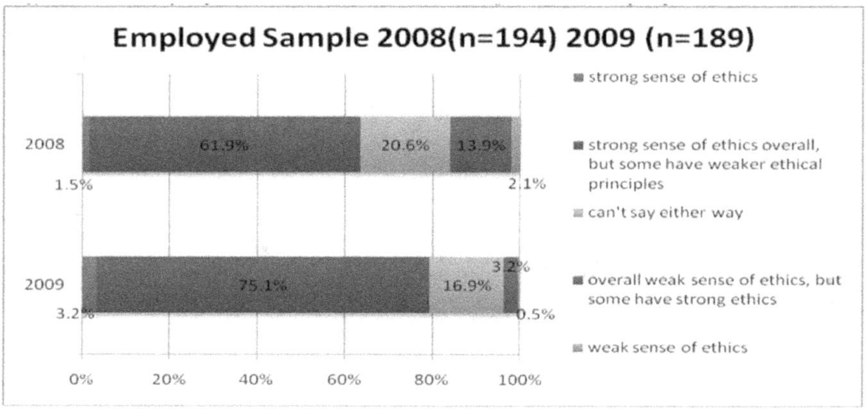

Figure 1.5. Employed Citizens' Evaluation of Public Employees. Created by the author with reference to the National Personnel Authority of Japan. Unofficial translation by Professor Fredrick Butcher. Question: "What are your current impressions regarding the ethical principles of civil servants?"

siders the personnel training programs of the National Personnel Authority. The NPA training experience of Mr. Ken Ishimaru (Mansfield Fellowship Program 2008–2009) is included as he participated at the Deputy Director level of training in 2009, which is the highest level facilitated by the NPA.

To begin, the Training and Development Policy Division and Training Administration Division of the Human Resources Bureau of the NPA along with four coordinate bureaus and the National Institute of Public Administration facilitate "Training and Human Resource Development" programs which are tailored for administrators depending upon their seniority in the system. In tune with the Nishio et al. (2010) Study Group's recommendations along with the 2009 Monitor Survey, the NPA's explanation for the goals and purposes of these training programs is to affect the manner of decision-making itself exercised by "defining the public interest that managers are expected to respect and serve" (Farazmand, 2001, 840). As the NPA's Training and Development Policy Division Clarifies (2010a, 2),

> Training programs are provided to personnel who are expected to play a central role in administrative management of each government ministry according to its position level (from officer level to director-general level). They aim to pursue the basic goals of enhancing the qualifications required for administrative officers, including development of a sense of mission as servants of the whole community based on wider vision, flexible mindset, rich sensitivity and high ethical standards, as well as building a relationship of mutual trust necessary to cooperate to implement measures with each other as public national employees.

Administrators are divided up for training first by the purpose of the training including: (1) training by level of position; (2) dispatching training (e.g., overseas placements, domestic fellowship placements in graduate programs, or private enterprise placements); (3) theme/target-specific training (e.g., mentor, mid-career, and female employee training); (4) instructor training (e.g., sexual harassment prevention, ethics, supervisory, business manner training). The focus here is on the first category of "training by level of position" which is program in theoretical training using a mixture of approaches. These training seminars target employees at the ministries based upon their respective career levels including (1) new employee, (2) third year employee follow up, (3) assistant division director level training, (4) division director level training, and (5) director-general level for executive administrators.

Reviewing the curriculum of the initial orientation for newly recruited civil servants is illustrative (see table 1.3 infra). Administrators' training times are broken into sections. In this particular case, trainees completed three days of training or twenty-three hours broken down into "Classification" (i.e., responsibility, public policy, fairness in public administration) and "Subject" (i.e., public officialdom, tradition, individual study, personnel and organizational management). The method used might include lectures, discussions, and actual practice. The focus of these approaches is to some extent classical and formulaic so as to provide then newest recruits the essential skills to begin working. At the next levels of seniority, however, the NPA implemented a host of innovative approaches as part of their human resource development efforts as well. Some of these approaches include:

1. "Mutual Development of Trainees: Implementation of 'participative curriculum' and 'camp style' training with emphasis on group discussions, exchange of views, and practical experiences."
2. "Intercommunion of People from Different Fields, Expansion of Views: Attendance of a variety of participants from the Cabinet Office and Ministries, Private sectors and Foreign Governments, etc."
3. "Essential Knowledge and Ideas for Administrative Officers: Lectures given by top-level professionals on contemporary issues and themes" along with exams and courses covering the "classics, ethics and civilization" among other things (NPA, 2010b, 12).

Three examples including (1) New Employee Administrative Training, (2) Administrative Training at the Assistant Division Director Level, and (3) Administrative Training at the Division Director Level are illustrative. First, the curriculum for new employee administrative training in FY2010 was five weeks in length and remarkable for its emphasis on hands-on experience through projects that facilitated direct interactions with the public. As table

Table 1.3. NPA New Recruit Public Administration Training

Classification	Subject	Method	Time
Thinking about the responsibility and duties of public servants	To be a public official	Lecture, discussion	1.5 hours
	Study of tradition	Lecture, discussion	2.5 hours
Thinking about public policy making from a variety of angles	Study of issues in public policy	Lecture, discussion	8.0 hours
	Individual study of public policy	Discussion	5.5 hours
Fairness in public administration and management	Personnel management, Organizational management	Lecture, practice	3.0 hours
Other	Introduction, Wrap up, Orientation		2.5 hours
Total: 23 hours			

Source: National Personnel Authority of Japan. Unofficial translation by Professor Frederick Butcher. Note: Public Administration Training (Assistant to Section Head Track).

1.3 reveals, in addition to regular classroom instruction, new employees also engaged in introspective exercises facilitated through site visits and involvement in local community projects with all of these activities culminating in a case study designed to encourage new employees to see problems as multidimensional and to seek solutions accordingly. This dovetails with recommendations 4 and 5 in particular of the Nishio et al. (2010) Study Group as the training incorporates introspective self-study and real world case study analysis to cultivate solution driven approaches to administrative decision-making.

Turning to the second example of a mid-level training program, at the beginning in the second month of his Mansfield fellowship James Miller (MFP 2003–2005) voluntarily participated in the NPA's Assistant Division Chief Training Course. The purpose of this training, as both testaments from other Mansfield fellows and the literature from the NPA also confirms, reinforces the view that "the Japanese regulatory process is fundamentally concerned with bureaucrats' *parens patrie* role as the enlightened protector of the public" (Miller 2006, 18). More specifically, the training is intended to "improve the skills necessary for the pursuit of their present or future duties, raise ethical standards, and contribute to the effective execution of public affairs" (Brosnan 1989, 11).

In addition to specifying how the duties of the Japanese bureaucrat are defined and delegated, this three week training course was directly relevant

Table 1.4. Curriculum for New Employee Administrative Training (FY2010)

1st week	Examine and consider the ideal public policies in a multifaceted way. • "Study on policy issues" (lectures, discussions, and preparation of recommendations on inter-ministry policy issues)
2nd week	Consider the ideal public administration through hands-on experience of nursing care etc. (about 130 facilities, such as welfare facilities for the elderly and facilities for the disabled)
3rd week	Consider the mission and responsibilities as servants of the whole community. • "National public service ethics," "classics," "sites of international administration," etc.
4th week	Consider the ideal public administration through hands-on experience of local administration (about 180 municipalities)
5th week	Again examine and consider the ideal public policies in a multifaceted way. • "Case study in administrative policies" (examples of adoption of the consumption tax, construction of an estuary barrage in the Nagara River, construction of Narita Airport, etc.)

Source: National Personnel Authority, 2010. "Training and Human Resource Development." Booklet created by the Training and Development Policy Division, National Personnel Authority, Government of Japan. 1-2-3 Kasumigaseki, Chiyoda-ku, Tokyo 100-8913, JAPAN: p. 3.

to Mosher's second variable providing the resources necessary for Miller to perform his responsibilities. Beyond the subject matter of the courses, which more than one Mansfield fellow described as academically rigorous including a concluding presentation by the fellow on a relevant policy topic, Miller discovered the practical value of becoming acquainted with one's colleagues in the upper echelons of the bureaucracy. As he clarifies,

> During the three-week course I met peers in every ministry of the government and formed important personal friendships. Attendance for part of the course required a week's stay at the NPA's training facility outside of Tokyo. . . . The relationships and work I did at the training continued to be a benefit throughout the year. [For instance,] I was able to be a resource even after the meetings through research, writing, and public speaking for many of the contacts I made at the meeting (Miller 2006, 8).

As such, the establishment of a professional network of contacts between public servants is also an important role that the NPA continues to play for Japanese bureaucrats. To expand further on this point, the NPA also facilitates relationships between bureaucrats and other powerful players such as elected officials, judicial officers, and private industry as well. More succinctly, one might note that this opens lines of communication between the executive, the regulators, and the regulated. As Mr. Miller's itinerary notes, the NPA sponsored a number of trips during his fellowship to locations such as the Toyota and Matsushita Headquarters (Miller 2006, 15). He further confirms this point noting that the ministry with which he was placed enjoyed close connections with both " . . . the academic and 'think-tank' policy space" (Miller 2006, 8). Consequently, administrators are encouraged to draw upon a reservoir of individuals with technical policy expertise despite their status as private entities (Miller 2006, 8).

> In an age where Japanese bureaucrats are obviously overworked and underappreciated, possibly millions of man hours of labor could be freed by simply requiring Japanese bureaucrats to incorporate greater public participation in policy development and public comment review and response, achieving a significant civil service reform at the same time (Miller 2006, 19).

Finally, turning to the highest level for Deputy Directors-General and Directors for large divisions, the NPA implements "a philosophy-based program dealing with a wide variety of classics and aims to cultivate a high sense of ethics required for public employees and foresight based on profound cultivation, through exchanging opinions and searching for the essence of excellent thought and the value of human beings" (NPA 2010b, 12). As Mr. Ken Ishimaru (MFP 2007–2009)[13] explains, an apt analogy is that participants come to see that the Japanese bureaucracy is like a wall of cobblestones rather than symmetrical bricks as is familiar in the western world. In other words, just like the ancient Japanese castles, the bureaucracy is constantly forming a sturdy wall out of cobblestones of varying shapes and sizes as personnel rotate from one office to another.

In contrast, U.S. federal agencies do not permit the same degree of interchangeability. Thus, the NPA's personnel training system embraces the bureaucracy's nuanced intercommunion, teaching flexibility and fit to keep the wall upright and reinforced. As such, at this level of training, the NPA gears activities towards systemic level questions. The question under evaluation in Ishimaru's case was how to enhance the bureaucracy's image in the eyes of the general public? As Ishimaru (2009, 17) explains,

> The National Personnel Agency (NPA) sponsored 207th Administrative Training (for deputy director class officials) entitled *"Thinking about the Future of Japan's Government/Civil Service ~ Regaining the Trust of the Public"* training in Tokyo and at the Iruma Training Facility was exceptional. The first week of training . . . consisted of attending seminars provided by some of Japan's leading authorities regarding the country's government, bureaucracy and civil service, as well as working in teams to come up with a project that would be presented to the entire group of 34 trainees. The lectures were all informative, while the team meetings were very stimulating. The final team project was to address issues and causes for losing the public's trust, and come

> up with ways to regain that trust and confidence. The second week of training included more seminars, along with the addition of the 34 trainees giving individual presentations. Since the theme of the entire training session was regaining the trust of the general public, my presentation was entitled "An Outsider's view of Japan's civil service system." My presentation went extremely well, and while most presentations ended during the allotted time or well before the required time, due to many questions and exchange of various views, mine went over the time limit by approximately 10 to 15 minutes.

Although Ishimaru avoided providing policy recommendations to safeguard against appearing excessively critical and/or arrogant, he nonetheless was asked to present on his overarching impressions of the differences between the U.S. and Japanese administrative systems. To that end he prepared a thorough presentation on his perceptions of the differences between administrative systems of the U.S. and Japan such as the greater emphasis on mentorship in the United States for new employees. As noted above, his audience was most receptive to his observations.

As the above three examples illustrate, at the start, middle, and height of their administrative careers, the NPA has designed programs which seek to shape the decision-making that public servants are expected to employ. For the newest "recruits," this is a three day basic skills course whereas for new "employees" this turns into an intensive five week training program to provide theoretical and practical guidance as they enter complex policy contexts. Finally, for mid-level administrators the focus is on building networks and cooperation as between agencies whereas for high-level division directors, the focus is on evaluating the system as a whole.

Given these carefully designed, assessed, and periodically administered administrative training tiers for administrators, one might presume that the NPA is in fact accomplishing its larger role as guardian of the public interest. However, when one considers the massive size of the bureaucracy as opposed to the numbers of personnel receiving this carefully crafted training experience (see tables 1.5–1.6), the relative inability to effectively implement training of this nature for all bureaucrats across the board becomes clear. Specifically, the NPA's staff has hovered at around 700 since 1997 to service what today remains over 55,000 general service employees even after the number of employees has been reduced dramatically over the past decade of reforms.

Thus, when the variety of official powers over the personnel system are added to these systemic responsibilities, the monumental nature of the task becomes clear.[14] Indeed, merely running sufficient events annually to be sure that all personnel in the mid-level assistant director level receive appropriate training is a significant task. Likewise, as administrators' often have heavy workloads, the NPA has frequently run more than one event and permitted personnel to select events that more closely match their substantive interests

Table 1.5. The Number of Government Employees and Total Personnel Cost

FY	Number	Total Cost (Billion Yen)
2006	58,900	5,408.6
2007	59,600	5,370.9
2008	59,100	5,325.2
2009	57,500	5,319.5
2010	56,400	5,179.5

Source: Figures courtesy of the International Affairs Division of the NPA. Prepared by the Ministry of Finance.

and dovetail with their work schedules. Since the start of the Mansfield Fellowship Program, this equates to over 3,800 employees at the mid-level of seniority alone (table 1.5). Likewise, although the training responsibilities of the NPA are substantial, as this project has revealed, the NPA is engaged in a host of other activities from facilitating the National Personnel Monitor Surveys and concomitant academic study groups to arranging the placements of the Mansfield Fellows in coordination with the MFP staff (see Appendix A for sample schedule) to managing the examination, onboarding, discipline, and retirement of employees.[15] Given these resource constraints, even after the 2014 reforms which transferred some of its responsibilities to the Cabinet Bureau of Personnel Affairs (CBPA) like career development, recruitment examinations, and staff number control (Mishima 2017, 9), the capacity for the NPA to run events of the caliber that both James Miller and Ken Ishimaru experienced, along with those that many other Mansfield Fellows, is remarkable. When coupled with the fact that the NPA has remained under significant political scrutiny along with finding enhanced duties to train a larger number of personnel on a tighter and tighter budget, it is no wonder that the NPA has been forced to shorten its training period from what was three weeks during Miller's MFP experience in 2004–2005, to one-and-a-half weeks during Ken Ishimaru's MFP stint in 2008–2009 to what is anticipated as a matter of days in the post-2014 period. Thus, it would seem that the trouble lies not in the mission statement nor in the NPA's understanding of its duties, but rather in the sheer lack of resources to effectively carry all of them out.

CHAPTER SUMMARY

As the foregoing discussion of the history, powers, and roles of the NPA reveal, this agency was established with an expansive mission to enhance the efficiency and responsiveness of the Japanese administrative system. Indeed,

Table 1.6. NPA Training Sessions: Number of Participants and Events Per Year at the Assistant Director Level

FY	Events Per Year	Number of Participants
1995	6	289
1996	6	309
1997	7	341
1998	7	320
1999	7	296
2000	7	293
2001	7	277
2002	7	253
2003	7	253
2004	6	233
2005	6	213
2006	6	199
2007	5	184
2008	5	185
2009	5	175
Subtotal (1995–2009)	*94*	*3,820*
Total*	***206***	***9,016***

Source: Courtesy of the International Affairs Division of the NPA. Sum of years 1995–2009 calculated by author.

* Total sum is counted from the first year of this training in 1965.

it is not overselling the point that with regard to responsiveness, the NPA was officially tasked with limiting bureaucratic drift and remains a key link between bureaucracy and democracy. It is the only agency tasked with assessing the administrative system as a whole. It is likewise the only agency with the powers and authority to intervene in the personnel affairs of all the ministries. This section has focused on the substantive aspects of the NPA approaches in tune with the recommendations of the Nishio et al. (2010) Study Group and supported through the national survey data available through the NPA's annual 2009 Monitor. To accomplish these goals, the NPA has developed tiered training programs with various methodologies tailored to the level of seniority of the participants. This attempts to concretize "the public interest that managers are expected to respect and serve" with the expectation that they will "control themselves" (Farazmand 2001, 840).

In practice, however, the NPA struggled to accomplish this larger role and thus to decrease bureaucratic drift by curtailing bureaucratic discretion and thus aligning agents with their duties as public servants. This is precisely why the NPA's impressive and expansive powers over the personnel system were dialed back in 2014. Changing the ways in which bureaucrats think and operate is a very challenging task. In the face of declining funding, limited personnel, and the trenchant political *milieu* in which the NPA was operating, it is perhaps only surprising that it took so long for the central government to transfer some of its powers to the Cabinet Bureau of Personnel Affairs. In short, there was clearly drift between the NPA and the Diet itself. Given the inability of the NPA to reduce bureaucratic drift and thereby constrain the discretion of administrative agents on a daily basis, the NPA had instead focused on enhancing the efficiency of the system from restructuring working hours to rethinking administrative leave and other aspects of the bureaucrats daily work lives.

The following chapters evaluate the consequences of these approaches. In other words, by focusing on the efficiency in public administration rather than addressing bureaucratic drift, responsive governance was the casualty. The following discussion, therefore, evaluates how despite the NPA's significant efforts to limit bureaucratic drift and directly control administrative discretion, the link between democracy and bureaucracy was severely stressed leading to the need for change.

NOTES

1. This subsection alone on the structure of the Japanese government draws in part from an undergraduate senior honors thesis penned by the author titled, "Bureaucratic Control of Japan: The Development and Functioning of a Traditional System" defended before a committee for credit in Independent Concentration: 99D at Brandies University on December 10th, 1999. See bibliography for reference. Importantly, proposing this classification of five perspectives was first undertaken by the author in this successfully defended thesis which received highest honors. A copy of this thesis is on permanent reserve at the Robert D. Farber University Archives and Special Collections, Brandeis University.

2. Importantly, since 1989, this trend seems to be changing. Specifically, in 2007 and 2009 under an LDP administration, and between 2010 and 2012 under a Democratic Party of Japan (DPJ) administration, the government of the day found its legislation largely blocked because the House of Councillors had an opposition majority. In particular, although the House of Representatives may ultimately triumph when legislative disagreements arise after specified time periods elapse, the lower house still requires a two-thirds majority to move forward. Thus, if the prime minister does not command a two-thirds majority in the lower house, the upper house may prove to be a significant roadblock. For constitutional legislation, this becomes an even greater challenge. Indeed, a prime minister like Shinzo Abe with constitutional goals can only realize changes to the 1947 Constitution with a two-thirds majority of members of both the House of Representatives and the House of Councillors (plus a simple majority in a referendum of the people) (Oyama 2015).

3. Officially, the Policy Research Council or *Jiminto Seimu Chosa-kai*.

4. Another key aspect of reelection that continues to remain salient is whether a candidate is a "dynastic legislator" meaning that the candidate is seeking to "inherit" the Diet seat,

campaign club, and "*kamban*" ("name recognition") from his father (Fukumoto 2011, 54). Indeed, studies show that sons in particular inherit 80–90 percent of the votes their predecessor obtained in the immediately preceding election (Fukumoto 2011, 54). In addition to dynastic legislators, incumbents in general have a distinct advantage in Japan. Even after the change to a single member district (SMD) electoral system in 1993 which enhanced the competitiveness of elections nationwide, incumbents from 1996–2005 succeeded in 61.5 percent of national elections down from 81.7 percent in 1983–1993 (Fukumoto 2011, 55).

5. MITI has since been renamed "METI" or the Ministry of Economy, Trade, and Industry.

6. This was hardly the case as a review of the documentary history reveals a rational decision-making assessment by the salient actors at the time such as at the Imperial Conference of November 5, 1941, of the necessity of war with the United States (Lu 1997, 425–35). In fact, there is some indication that the decision to bomb Pearl Harbor was made in spite of the recognition that the outcome would probably be disastrous for Japan. The choice to proceed was rather envisioned as the only reasonable alternative to a continued economic stranglehold by the United States which had choked off oil exports to the resource poor island. While not typically envisioned in this light, U.S. entry into World War II was thus instigated over oil and resource dependency considerations and not the machinations of irrational or mystic beings across the Pacific.

7. This linkage between the need to govern Japan's new democracy through the extant bureaucracy is a well settled historical fact. See Justin Williams (1968), Dolores Brosnan (1989), B. C. Koh (1989), and, more recently, M. Laxmikanth (2009).

8. Blaine Hoover returned the next year in 1949 to the United States to head the Commission on the Organization of the Executive Branch (a.k.a., the "First Hoover Commission") to help solidify the contracted role of the Civil Service Commission, which ironically looked a great deal more like the watered down version of the NPA favored by Japan's Diet members that he had so vehemently opposed during his time with the SCAP (See OPM 2011, 17 for a description of the some of the personnel recommendations produced by Hoover's hand.)

9. It is unclear how many of these activities will continue after the 2014 reform.

10. The NPA is also tasked with measuring and evaluating the performance of administrators in Japan in tune with Mosher's third component of responsibility. This process is very complex and has changed significantly over the past decade. Finally, the NPA is responsible for ascertaining violations of the National Civil Service Ethics Law through a formal adjudicative process, which conforms to Mosher's fourth component requiring sanctioning for failing to comply with those duties within specific ethical parameters (NPA 2010c).

11. As a point of comparison, it seems most unlikely that over 50 percent of Americans are familiar with the U.S. Office of Personnel Management and might articulate its responsibilities.

12. One such slogan that does not translate well into English exhorts officials to "Start from yourself to behave yourself, which you can be proud of" (NPA 2010c, 179).

13. Ishimaru is a U.S. federal agent with significant experience in operations, investigations, analysis and U.S.–Japan bilateral relations. He possesses native level proficiency in Japanese, as per the U.S. Department of State's Foreign Service Institute requirements.

14. Significantly, the total number of public service employees in Japan as of 2016 was 640,000 of which 275,000 were subject to the NPA's annual Recommendation to the Prime Minister under the Remuneration Act (NPA 2016, 2).

15. Jay Biggs (2010, Personal Communication), for example, described in depth the participation of a Chinese delegation during his NPA training experience which permitted the sharing of opinions across cultural lines as well. Survey data that the NPA maintains from Japanese administrators who have participated in the training projects is also positive. In FY2010, for example, results of a questionnaire distributed to new employees who had completed training indicated: 163 persons (40 percent) fully satisfied; 214 persons (52 percent) almost satisfied; 23 persons (6 percent) neither satisfied nor dissatisfied; 8 persons (2 percent) somewhat dissatisfied; and 0 persons (0 percent) dissatisfied (NPA 2010a, 3). An equally high level of satisfaction was recorded amongst participants at the mid-level training in FY2009 completed at the NPA Kinki Regional Bureau. Moreover, anecdotal evidence from both groups indicates that at least six participants whose comments were included in the booklet better understood their responsibilities as public servants. For instance, a new employee in 2010 writes of the hands-on

training experience during the session, "In local communities and nursing-care facilities, there are many people who ha[d] contact with national public employees for the first time. Whether people's confidence in the national government is increased or undermined depends on trainees' behaviors. I thought it was important to participate in training with this awareness in mind" (NPA 2010a, 3). Likewise, from the mid-level officer training a year earlier, a Japanese participant opined, "I was able to rediscover the basis for the concept that national public employees were servants of the whole community. I hope to carry out my duties with this awareness in mind in the future" (NPA 2010a, 4).

Chapter Two

Case Study 1

Ministry of International Affairs and Communication (MIC)

INTRODUCTION

The first ministry evaluated is the Ministry of Internal Affairs and Communications (MIC). The following first provides background on MIC before turning to the experiences of James Miller and Yukiko Ellis who both completed their Mansfield fellowships at MIC. Although MIC is involved in highly technical and scientifically complex responsibilities in the two policy areas evaluated including radio wave regulations and population statistics, it acts in a multi-actor policy network of shifting alliances. To reiterate, the question is whether *competition by multiple principals creates opportunities for bureaucratic drift*. In turn, we hypothesize that *policy settings including multiple principals (independent variable) are more likely to manifest bureaucratic drift (dependent variable). At the same time, policy settings excluding multiple principals (independent variable) are more likely to manifest less bureaucratic drift (dependent variable)*. Once again, *variation in agent discretion is the critical effect of the independent variable (i.e., number of principals) on the dependent variable (bureaucratic drift)*. Lastly, *"discretion" is "administrative decision-making absent directives from political principals."*

Accordingly, the first example presented below confirms that due to the presence of multiple principals in this policy network, enhanced bureaucratic drift is evidenced. As predicted, MIC officials have enjoyed significant discretion in policy initiatives. MIC administrators have been able to adjust their policy preferences and internal rules to changing circumstances and situations in ways that are advantageous to MIC. This is true even though this has meant that MIC frequently changes course as the principals in its network and the larger policy environment shift. Importantly, the personal preferences of administrative agents are frequently driving policy decisions rather than the informed evaluation of data or careful reliance on standard operating procedures. At the same time, as confirmed by the second example below, in the absence of multiple principals, less bureaucratic drift is apparent. In turn, agents are vested with less discretionary authority.

By way of background, MIC deals with highly specialized policy fields that would seem to require administrators with significant expertise and training in statistics along with technical savvy. Likewise, its various projects oftentimes implicate both local and centralized government functions as they touch upon national affairs. As MIC has described its primary responsibilities,

> The Ministry of Internal Affairs and Communications (MIC) has jurisdiction over various systems involved in the fundamental framework of the nation. . . . MIC's administrative functions are closely involved in the infrastructure of people's lives. . . . MIC is committed to promoting administrative reforms, disclosing administrative information, advancing regional decentralization, ensuring stable sources of local tax, promoting municipal mergers, activating regional economies, securing people's safety, and building increasingly sophisticated systems for information and communications. All this [is] done by . . . segregating the duties of the national government, local governments, and the private sector.[1]

Examples of MIC's bureaus provide an indication of the Ministry's expertise in heavy data driven responsibilities along with technologically challenging projects. First, it supervises a Personnel and Pension Bureau for over 80,000 public employees. Second, it is responsible for a variety of substantive policy areas as revealed by the titles of its bureaus such as the Administrative Evaluation Bureau, Local Administration Bureau, Local Public Finance Bureau, Local Tax Bureau, Global Information and Communications Technology (ICT) Strategy Bureau, Information and Communications Bureau, Telecommunications Bureau, and Statistics Bureau. To take one project supervised by the Local Administration (LA) Bureau detailed in MIC's *2010 Whitepaper*, the bureau identified three national issues including (1) health and medicine, (2) education and labor, and (3) lifestyle and living that would benefit from the significant expansion of E-government services (MIC 2010, 2). In the first category of health and medicine, the LA Bureau oversaw the creation of an E-government health management Web service site "allowing users to manage their own health information including personal data (height and weight . . .) and medical history and their average amounts of daily exercise . . ." (MIC 2010, 2). This was in order to "obtain advice tailored to their physical condition and information on examinations being conducted in nearby municipalities, helping them to prevent the outbreak or worsening of illnesses" (MIC 2010, 2). As a number of these projects touch upon national personnel issues, frequently MIC has worked closely with the National Personnel Authority (NPA) to achieve implementation (MIC 2010).

While one might presume that MIC's policy positions are dictated by its data driven activities, however, MIC is actually a lively policy actor frequently shifting alliances as needed to safeguard its power. As Kenji Suzuki (2006a, 182) explains, "Telecommunications policy requires great technological expertise, but this does not mean that the policy stands on purely technological grounds. On the contrary, it seems to be quite vulnerable to political intervention."

This is possible because, as Taplin and Wakui (2006, 10) elucidate, MIC operates in a policy network without an "iron triangle, meaning a stable and closed relationship among industries, ministers and politicians, which is seen as a feature of the Japanese political and economic system in other sectors." Iron triangles are critical for providing not only stable relationships, but a stable policy orientation over the long-term since these groupings form relatively coherent policy goals. Rather, MIC operates in a swirl of shifting alliances Suzuki (2006a, 184) calls "cliques" which only differs from Paul Sabatier's (1999) "advocacy coalition" in that "a clique may also be formed regardless of policy ideas."

Suzuki (2006a) continues this story with an extraordinarily detailed account of this policy network beginning in 1980 with the founding of the Ministry of Posts and Telecommunications (MPT) to its eventual merger with other ministries and agencies in 2003 to form MIC, and finally to its revised policy network and position during the Koizumi administration in 2005. Suzuki begins with the emergence of the MPT in 1980 as a "policy" rather than merely "operational" ministry with enhanced prestige—particularly in the eyes of competing ministries such as MOF and MITI—accompanying its new role (Suzuki 2006a, 186). Among its first major policy initiatives was the question of whether to privatize the Nippon Telegraph and Telephone (NTT) "to secure market competition in the telecommunications sector" (Suzuki 2006a, 183–185). Japan boasts a telecommunications industry and network that is "one of the largest-scale in the world . . ." (Taplin and Wakui 2006, 1). The MPT initially opposed privatization as it was fearful of losing public control over the massive NTT corporation and thus its job if the telecommunications market was to actually become competitive (Suzuki 2006a, 187, 200). However, two decades later after MPT had become MIC, it completely reversed course choosing instead to support the NTT (Suzuki 2006a, 200).

Without going step-by-step through Suzuki's careful delineation of each and every actor in the policy network and the details of MIC's changing policy posture, the importance of his overall message is clear. The telecommunications policy arena is marked by significant involvement of multiple principals that form ever shifting cliques of interests. These included competing policy tribes ("*zoku*") in the Diet both siding with the MPT (*Teishin-zoku*) and its ministerial rivals including the Ministry of Finance (*Okura-zoku*); advisory bodies such as the *Rincho* appointed by prime ministers from Nakasone to Koizumi due to significant public attention; business interests represented by such figures as Toshio Doko, famous both as a businessman and former chair of the Japan Federation of Economic Organizations (*Keidanren*); the NTT trade union (*Zendentsu*); and powerful personalities who came and went from the infamously corrupt Tanaka Kakuei in the 1980s, to the deputy secretary-general of the Liberal Democratic Party (LDP), Hiromu Nonaka, in the 1990s to 2003, to Koizumi until 2006 (Suzuki 2006a).

Although Suzuki posits multiple rationales for MPT/MIC's change of attitude on the question of privatization, much evidence points to MPT/MIC's desire to retain its "centrality" in the policy network. As Suzuki (2006a, 184) explains, "[c]entrality denotes the extent to which each stakeholder lies near to the centre of the network . . . [because] the most central stakeholder has the largest influence on the decision of the policy output." As such, rather than basing its standpoint on technical evaluation of telecommunications policy, the MPT/MIC sought to retain their independence from the Ministry of Economy Trade and Industry (METI)—a long-term foe with disparate regulatory desires—and to avoid the dilution of its powerful regulatory role in telecommunications policy (Suzuki 2006a, 200).

To further evaluate how MIC operates in complex policy networks with and without multiple principals, the experiences of two Mansfield Fellows are woven together including James J. Miller and Yukiko Ellis. Their fellowship placements with MIC—provide critical case studies to asses both hypotheses proposed. In the first case, Miller's observations of the telecommunications policy network is remarkable for the significant participation of principals as noted by Suzuki (2006a) in the introduction to MIC above. In contrast, the lesser known activities of the Statistics Bureau of Japan at which Ellis completed her Fellowship, is remarkable for the absence of significant input by principals. Both cases thus provide two disparate policy contexts that MIC operates within that permit an evaluation of the research question. Specifically, with regard to the Telecommunications Bureau, the *presence* of multiple principals (independent variable) is predicted to enhance bureaucratic drift (dependent variable) with the effect of enhanced discretionary authority of bureaucrats. On the other hand, the *absence* of multiple principals (independent variable) in the administration of the census by the Statistics Bureau is predicted to decrease bureaucratic drift (dependent variable) with the effect of decreased bureaucratic discretion.

JAMES J. MILLER: MINISTRY OF INTERNAL AFFAIRS AND COMMUNICATIONS

James J. Miller, who completed his Mansfield Fellowship (2004–2006) with MIC with the Telecommunications Bureau, Radio Policy Division, provides an instructive case study where multiple principals are involved. Specifically, this case affirms that the presence of multiple principals increases bureaucratic drift with the effect of enhanced administrative discretion.

By way of background, at the time this data was collected, Miller was a senior attorney for the Federal Communications Commission (FCC) and an adjunct professor of law at American University, Washington College of Law. He possesses outstanding Japanese language fluency allowing him to "blend in" and "to be as Japanese as [he] choose[s] to be" (Miller 2006, 7). His three fellowship objectives included to:

1. "Develop a strong understanding of fundamental legal, regulatory, and political structure of policy making."
2. "Develop a strong understanding of the Japanese approach to regulating new technologies."
3. "Develop a strong understanding of the Japanese approach to communications regulation" (Miller 2006, 3).

To this end, he spent six months with MIC's Telecommunications Bureau, Radio and IT Planning Group. Near the end of his placement with MIC, he evaluated various legal and policy reforms under review on topics regarding transparency and comparisons with the U.S. regulatory structure (Miller 2006, 10). The Telecommunications Bureau also has a clearly demarcated set of policy responsibilities including,

> to promote competition in the telecommunications business, the development of a secure and reliable environment for using information-communications infrastructure, efficient use of radio waves and establishment of a globally-advanced wireless broadband environment to handle such changes.[2]

He delivered numerous presentations during this placement on topics ranging from cognitive radio and radio environment monitoring, to relationship banking and tax benefits for Japanese legal entities. For instance, to become briefed on spectrum policy, he visited the Radio Environment Monitoring and Enforcement Office, conducted an inspection tour of the Kanto Region General Communications Bureau to assess their enforcement and monitoring facilities, and presented on Cognitive Radio Technology and FCC Wireless Policy Efforts at the National Institute of Information and Communications Technology (NICT) Yokohama Research Park. Notably, these were just a few of his activities in January of 2006 alone. His itinerary for each of the preceding four months from September to December was equally impressive. Subsequently, Miller completed an NPA training program, along with placements with four courts including the Tokyo District and High Courts, and the METI. He concluded with a presentation on "The '*Doshusei*' Reform Bills and Comparative Federalism: Political and Legal Theory" along with meeting the Crown Prince and Princess of Japan (Miller 2006).

Regulatory Backdrop

To set the backdrop of the larger regulatory context, Miller arrived at MIC shortly after the "Law to Revise Certain Aspects of the Administrative Procedures Act" was effectuated by cabinet order in April of 2006. The purpose of the statute was to adopt a "public comment" process allowing greater input from the public in the policymaking of all the ministries (Miller 2006, 16). This was consistent with ongoing reform efforts to push the ministries to be more responsive to various interest groups in civil society.

However, while certainly a new step, the law lacked certain features found in the U.S. Administrative Procedures Act (APA) at the time which watered down its potency from Miller's standpoint. In the U.S. regulatory context, Miller perceives the public comment process is critical since it

> empowers the public to directly impact the substance of the government policies. Certainly no agency is required to entertain frivolous petitions, but where the process demonstrates there is reason to act on a petition, an agency is required to vet the ideas. Imposing this legal burden on an agency, provides a mechanism for the public to vet new ideas and respond to them. A very important effect of this legal burden is to deny agencies the opportunity to ignore new ideas when they come forward (Miller 2006, 18).

In contrast, Japanese regulators are skeptical of this reliance on market-based interests to drive policy innovation. As Miller (2006, 28) clarifies particularly with regard to MIC,

> [T]he Japanese approach to licensing and allocations is [to] follow international activity and vet with key players in various research committees and advisor[y] boards what new technologies should be employed. . . . After having 'refined' the discussion, Japanese regulators will conduct a rulemaking. Some Japanese regulators view with significant discomfort (if not hostility) the 'U.S approach' marked by volatile market changes and the frequent entry and exit of players (sometimes through gasp, bankruptcy), technologies, and service offerings.

Thus, even prior to this reform, the telecommunications "regulatory process [was] fundamentally concerned with bureaucrats' *parens patrie* role as the enlightened protector of the public" (Miller, 2006, 18). Thus, the new Act in Japan provided the opportunity for the first time for Japanese agencies to "shift . . . some portion of the burden of formulation and review of complex policy matters to interested parties" (Miller 2006, 19).

As Miller (2006, 18) explains, however, even though the process would seem to have represented an opening up to a new set of principles, the newly won power of the public was more illusory. In fact, the Act continued to privilege the agents in this multiple principal policy network. Specifically, the U.S. context affirmatively requires the *consideration* of viable ideas whereas the Japanese process only requires the *collection* of those ideas. Thus, joint data *collection*, but not joint data *evaluation* is a feature of how Japan's ministries interact with the public. As a result, these rules permit administrators to preferentially choose whether to look into new ideas from outside the ministry.

Public Comment and Telecommunications Policy

Miller then observed how this public comment issue played out in the context of telecommunications policy formulation and implementation at the Ministry of Communications. Miller's (2006, 23) specific professional interests were in Software Defined Radio (SDR) along with Cognitive Radio and related technologies. As Miller envisions the salient regulatory issues, the basic problem was how to allow innovations in the underlying technology while preventing harmful interference of communications by unlicensed and unregulated radio devices (Miller 2006, 23). As Miller (2006, 24) explains specifically with regard to SDR, "certification can be summed up as a problem of who and under what circumstances should a user be able to modify the software running on a radio."

Miller observed that the differing answers to that question crafted by the U.S. FCC and the Japanese MIC were due to their disparate underlying goals. The FCC when considering SDR certification explained, "We neither wish to have our processes inadvertently be a barrier to the development and deployment of these technologies nor wish to permit the widespread deployment of radios easily susceptible of being misused to cause harmful interference to others" (Miller 2006, 24). In other words, the twin goals were flexibility and conformity. In marked contrast, the Japanese approach to SDR emphasized a tougher regulatory stick. Specifically, MIC's mission statement explained that "[t]he Ministry of Communications referencing the submitted comments in this proceeding, will proceed by monitoring the direction of [research and development] for SDR and international activity" (Miller 2006, 24). As such, MIC undertook a much more "cautious" direct oversight role of SDR technology meaning that while research and development would be encouraged, until it had been fully vetted, SDR would not be permitted in Japan (Miller

2006, 26). As Miller (2006, 29) explains, this is common in Japanese policy-making beyond the telecommunications arena since "[i]t is common for Japanese regulators to fund research until a technology matures sufficiently to go forward in the regulatory context."

Interestingly, the outcome for the FCC and MIC was that each ended up atop very different regulatory structures. On the one hand, the FCC took a proactive stance on SDR by forming a Technological Advisory Council (TAC) to ascertain the state of the art and possible directions for SDR (Miller 2006, 25). This was in tune with its embrace of the "unlicensed" or "commons-based spectrum regulatory environment" (Miller 2006, 27). In short, this unlicensed approach balanced flexibility and protections which meant apprising developers of the concerns of harmful interference and allowing them to build safeguards directly into the devices to ensure their compliance with the rules (Miller 2006, 27). To ensure this process was enforceable at the end of the day, the FCC retained the power to impose strict penalties on violators who failed to appropriately design their products (Miller 2006). Miller (2006, 27) concludes in the context of Cognitive Radio, "With the confidence that devices will comply with whatever technical rules are adopted before they enter the stream of commerce, rules can be crafted . . . to bring greater value to the public while preventing harmful interference." As a result, innovators were rewarded, violators punished, and the FCC received lively commentary on a variety of innovative and interesting proposals (Miller 2006, 26).

This example stands in stark contrast to MIC's cautious oversight role with a less dynamic and more restrictive public comment process. As noted, rather than adopting a flexible approach combining "pre-checks" of radios with a "commons-based" regulatory environment, MIC began by establishing a study group of experts to test SDR and evaluate conformity issues (Miller 2006, 27). It then sought a public comment process and after two years of evaluating proposals on security, certification, and other issues, it released proposed changes and rules to permit SDR use in April of 2005 in Japan. However, Miller clarifies that "MIC's stated intent [was] to follow the state of development . . . and the overall pace of the inquiries reveal a cautious stance on SDR" (Miller 2006, 28).

While this may seem to be a slight difference from the unlicensed commons-based approach by the FCC, the outcome tells a very different story. Unlike the FCC's pre-check system, MIC's approach meant that it would need to become involved in

> costly and burdensome field enforcement employing a highly-advanced spectrum monitoring network (DEURAS) to track "unlicensed" illegal emissions in real time. It also expends considerable funds on outreach which is why you

> may see posters with "stop the illegal spectrum emissions" around the Tokyo subway system (Miller 2006, 28).

In the absence of the FCC's regulatory authority to intercept potentially harmful radios before they enter the market stream, MIC is saddled with an extremely complicated oversight responsibility for cracking down on violations in progress (Miller 2006, 28).

Thus, rather than a reduction of bureaucratic discretion, the adoption of a public comment process by the ministries ultimately dramatically increased bureaucratic discretion to carry out its mission. Indeed, as this case of MIC's adaption of the public comment process to SDR and Cognitive Radio clarifies, increasing the number of principals involved resulted in extraordinary bureaucratic drift as the discretionary powers of MIC rose rather than the involvement of the general public (i.e., principals) as was intended. Likewise, in the absence of consensus as to the best technological approaches to achieve regulators' safety aims as is pursued in the United States's commons-based approach, regulators found themselves in an "awkward" position as maverick firms began filing administrative lawsuits against MIC (Miller 2006, 29). Rather than being bound as Weberian cogs into the machinery of government, MIC administrators have literally enshrined in the rule the expectation that their preferences trump competing policy proposals from principals and specifically the informed public.

As such, this case affirms the expectations of the research question that the involvement of multiple principals enhances opportunities for bureaucratic drift. Telecommunications is a lively policy setting in Japan including multiple principals. As even more principals were added to this policy network, agents drifted away from the intended purpose of including the general public more directly in policymaking. Rather, agents seized the opportunity to enhance their own discretionary authority.

YUKIKO ELLIS: MINISTRY OF INTERNAL AFFAIRS AND COMMUNICATIONS, STATISTICS BUREAU OF JAPAN, POPULATION CENSUS DIVISION

This next case study provides an excellent counterpoint to the preceding example in that multiple principals are not involved in this policy network. Specifically, this case affirms that the absence of multiple principals decreases bureaucratic drift with the effect of more constrained administrative discretion.

Yukiko Ellis, a mathematical statistician in the Bureau of the Census for the U.S. Department of Commerce, completed a Mansfield Fellowship (2002–2004) with the Statistics Bureau of Japan (SBJ). Specifically, her placement was in the Statistical Survey Department (Population Census Division) at what was then the Ministry of Public Management, Home Affairs, Posts and Telecommunications. The ministry assumed its new name as the MIC the year after her fellowship in 2005.

Ellis has an extensive background in this policy context. At the U.S. Census Bureau, Ellis has worked on the Post-Enumeration Survey for the 1990 Decennial Census, cognitive studies of the under-count enumeration, surveys and censuses of governments, and the 2007 and 2012 Economic Census. She has also contributed noted studies to the field. For instance, her work on the 1990 Decennial Census has been made available to the general public by the U.S. Census Bureau (e.g., see Ellis 1990) and was referenced during her Fellowship in personal conversation by Professor Kojima of the National Institute of Population and Social Security in Japan (Ellis 2004a). Ellis holds her BA in mathematics from the State University of New York at Binghamton, MPH in biometry from Yale University, and MS in biostatistics from the University of North Carolina, Chapel Hill. She is also fluent in Japanese as a native speaker. Since the completion of her Mansfield Experience, she has maintained close ties with the SBJ returning every year to renew contacts and continue to facilitate relations with the United States Census Bureau and the SBJ.

As this reveals, Ellis' expertise is firmly established and she brought very strong credentials to her Mansfield Fellowship placement at MIC. Specifically, as per her own contemporaneous assessment in her Mansfield Final Report (Ellis 2004a), Ellis accomplished the first two of the following four Mansfield Proposal Objectives:

1. "Obtain[ing] an overall picture of census taking including [the] Populaion Census Division's organizational structure, planning, design, and its project management style."
2. "Learn[ing] how [the] Population Census Division handles . . . survey operations, specifically that of data collection."
3. "Gain[ing] understanding of non-sampling errors, especially those arising from coverage error (i.e., under- and over-coverage) in the census frame that consists of a list of ho[u]sing units and from [the] non-response problem."
4. "Observ[ing] how the collected data are evaluated and analyzed."

To do so, required significant access to this policy setting. Ellis's firsthand observer data stems from working daily in the SBJ, attending a multiplicity of conferences and meetings, conducting numerous one-on-one interviews with Japanese civil servants, completing tours of various facilities, in situ reviews of documentary materials and reports, participating in the ten-day 178th NPA Administrative Training for mid-level managers, attending staff training sessions at the Statistical Research and Training Institute (SRTI), and participating in a variety of meetings with other government agencies at both the central and local levels of administration (Ellis 2004a).

In addition to her impressive personal credentials and ability to smoothly interface with this policy setting, she also had the added cachet of representing the U.S. Census Bureau, which enjoyed a powerful reputation in Japan as a centralized and professional administration. Concomitantly, the SBJ at the time of her Fellowship was very cosmopolitan as it was heavily engaged in the international policy context. For instance, MIC hosted two international conferences in December of 2003 including the Meeting of the Heads of National Statistical Offices of East Asian Countries and the 21st Population Census Conference, which was attended by administrators from 20 countries and 5 international organizations (Ellis 2004b). As such, Ellis was warmly welcomed into the SBJ as the staff was "genuinely excited" for the opportunity to work with her (Ellis 2004a). Thus, Ellis is a credible and strong source for evaluating MIC's policy network as ascertaining both the "organizational structure" and the "project management style" of the SBJ were specified objectives of her fellowship.

The following data was garnered through her responses to a written survey, a personal interview, review of detailed monthly and final reports to the Mansfield Foundation, subsequent communications to improve the accuracy of the following rendition, and other supporting documentation as appropriate. Nonetheless, while the following analysis presents her observations, ultimately this assessment belongs to the author as is true of all five case studies presented in chapters 2 through 6. This evaluation first begins with Ellis's observations at the SBJ before turning to her fieldwork in ten local prefectures.

SBJ: Population Census Division

This section considers Ellis's efforts to accomplish her first objective to "[o]btain an overall picture of census taking including [the] Population Census Division's organizational structure, planning, design, and its project management style" (Ellis 2004a). The specific Population Census Division with which Ellis completed a large portion of her Fellowship experience has well-defined responsibilities including

> executing and tabulating the Population Census, and the Housing and Land Survey. It also compiles current population estimates and internal migration statistics. The Labor Force Statistics Office in this Division is responsible for statistical surveys on employment and unemployment such as the Labor Force Survey and the Employment Status Survey, and the Survey on Time Use and Leisure Activities.[3]

At first glance, therefore, one might presume that the policy context was relatively apolitical with a premium placed on technical knowhow. In operation, however, Ellis discovered that the SBJ was faced with a variety of challenges, which hindered a purely data driven approach to problem solving. All of those challenges are built into the procedural framework of the public administration system itself.

First, the staff at the SBJ is not as well-trained in statistical analysis and data interpretation as one might expect (Ellis 2004b). Importantly, only the upper echelons of MIC are comprised of examination Level 1 employees who are more likely to have training in statistics and methodology. Instead, most of the employees who staff MIC receive on-the-job training having passed the more general Level 2 and 3 examinations. These public service entrance exams only require a college or high school diploma respectively. Thus, a Japanese administrator at the SBJ might not have completed college coursework in statistics. Consequently, Ellis found that much of the SBJ staff is comprised of "generalists" who are "almost clerical."[4] As a result, the toughest statistical questions are often reserved for a special Statistics Council and meetings with academia (Ellis 2011, personal communication).[5]

Second, not only is the staff of the SBA populated with generalists, but also those generalists are often given insufficient time to learn the ropes to offset their lack of training. Instead, all three levels of employees are routinely rotated in and out of their positions every two to three years. As Ellis (2011) explains,

> Everyone in all levels experiences periodic personnel shifts. Those in Level I tend to move within SBJ or their assignments outside of SBJ tend to be shorter. They all learn their new responsibilities on the job, rather than getting specific training, when they get assigned to new locations.

Consequently, the most frequently used expression in the workplace during Ellis's fellowship placement between administrators whether in the office or on the telephone was "please teach me" (Ellis 2004a). Again, the rotation system is the modus operandi for all government agencies in the Japanese system of public administration. Therefore, prior to launching new policy initiatives such as population surveys, Ellis discovered that the Statistics Bureau of Japan was forced to undertake expensive training sessions for both local and central government staff. As Ellis (2009, personal communication) explains the consequences,

> "So, you can't nurture the staff. You constantly spend so much time in training a new batch of people." "In statistics, we need experts. We need trained people. We need to be able to retain some of the same people." "[The top management] realize[s] it."

Third, these challenges were compounded by the size of the agency relative to its responsibilities. By way of comparison, Ellis discovered that MIC is a small entity with only approximately 600 employees in contrast with the approximately 4,500 employed by the U.S. Census Bureau at the time. By way of example, Ellis notes that in population census planning, production, maps, and operations, only 15 people at the SBJ were tasked with a job that 150 people would have handled in the United States (Ellis 2004b). Staffing constraints meant that the SBJ was also technologically challenged. For instance, ironically, although it was considering how to expand E-government services to the citizenry, the SBJ did not launch a Web-based system of data collection until 2004 (Ellis 2004b). Although they have done well since then to adapt to the new technology, their small staff size continues to present challenges for fully leveraging the system.[6]

Fourth, as this also implies, the small staff have very heavy workloads that seem to hinder more comprehensive data collection and analysis (Ellis 2009, personal communication). In fact, this problem had become a focal point of NPA efforts during Ellis's fellowship. On a number of occasions, Ellis recounts that a recorded message came over the intercom from the NPA to the staff of MIC encouraging everyone to abide by their regular work day dismissal time of 5:00 pm rather than working long overtime hours. This is in accord with continually updated national labor standards implemented by the NPA that have sought to encourage a healthier and reasonably balanced work life for public administrators. This remains an official goal of the NPA's "Realization of a Comfortable Working Environment" initiative which emphasizes "Measures to Support the Balance of a Working and Family Life." Provisions included providing additional flexibility for shortening the work day, to child care time and participation leave, maternity leave, nursing leave, and restrictions on late night work from 10:00 pm to 5:00 am, to additional sick leave and "special leave for summer vacation, marriage, before and after birth, mourning," and other reasons (NPA 2010b, 19). Even so, since they did not want to inconvenience their coworkers by failing to meet deadlines, MIC employees generally declined the invitations of the NPA broadcasts to leave early because of their heavy workloads.

All four of these types of constraints including insufficient methodological training, the rotation system, staffing shortages, and heavy workloads thus interfere with exclusively data driven decision-making. The careful assessment of data presumes expert administrators who have garnered sufficient knowledge over the course of their careers to handle workloads within reasonable limits. Upon inquiring why this system is permitted to continue perpetuating the same unhelpful procedures such as training and retraining every few years in a seemingly endless cycle amidst staff shortages and heavy workloads, Ellis's postulated that any other approach would likely require a macro level change across Japan's system of public administration reaching beyond the MIC. For instance, all the ministries would be forced to alter their training procedures and hiring expectations for different levels of employees. This would render any such change much more challenging to facilitate.[7] As this evaluation reveals, problem solving in this policy setting is significantly constrained by the structural framework and extant standard operating procedures.

Ten Prefectures: Fieldwork

The foregoing discussion thus clarifies the limitations on discretionary decision-making at the SBJ. The following reviews a specific policy context to permits the evaluation of the postulated effect of decreased discretionary authority that the absence of multiple principals (independent variable) has on bureaucratic drift (dependent variable). This example relates to Ellis's second and third objectives including (1) "[l]earn[ing] how [the] Population Census Division handles . . . survey operations, specifically that of data collection" and (2) "[g]ain[ing] understanding of non-sampling errors" (Ellis 2004b, see above). In particular, focusing exclusively on MIC misses the key point that this policy arena is comparatively decentralized vis-à-vis the United States or Canada (Ellis 2009, recorded telephone interview). Rather much power over policy implementation is devolved to the forty-seven prefectural governments to carry out census and other statistical data collection. This foundational observation is critical as it distinguishes this policy setting not only from other states, but also from the other Japanese policy contexts considered in chapters 3 through 6.

To begin, one of the most frequent observations of Japan's system of public administration is that the central government in Kasumigaseki, Tokyo's government district in which the ministries are housed, holds all the cards. This is oftentimes true in various policy contexts such as the historically observed role of MITI (now METI) in shaping trade and industrial policy (e.g., Johnson 1995) and MOF in crafting finance policy (e.g., Hartcher 1998) in the post–World War II era. In this domain, however, the situation is reversed. Japan's central government must rely upon the assistance of the local governments to complete nationwide statistical surveys. Thus, the SBJ operates in an unusually decentralized policy arena.

As a result, in addition to the constraints listed above on the staff of the agency, the SBJ faces many challenges when interfacing with local governments. Ellis once again provides valuable insight into this decentralized sharing of power. The following focuses on her travels with fourteen staff members in the Population Census Division to approximately ten prefectures in February of 2004 (Ellis 2004b). The general purpose of the fieldwork was to exchange opinions with prefectural and local level officials on implementing the upcoming nationwide census (Ellis 2004b). The specific objectives included (1) the delivery method of the census questionnaires, (2) establishment of an examiner position to review questionnaires to bypass prefectural level reviews, and (3) the evaluation of prefectural hiring practices of interviewers for the census (Ellis 2004b).

To focus solely on the first issue of the delivery method of census questionnaires, the complexities which inhere in the process become immediately apparent. First, one of the main topics of interest that Ellis initially and ultimately unsuccessfully sought to numerically evaluate in Japan as her third placement objective (see supra) was the quality of the surveys conducted in terms of sampling errors and other factors (Ellis 2004a; Ellis 2004b). When she sought to conduct a formal study into these issues upon first arriving in Japan in October of 2003, she had been apprised that data on survey quality such as item and unit nonresponse data from respondents were not collected "because, for one, the response rate in the Japanese population was so good that there was no need for such data" (Ellis 2004b). Second, although this is standard practice in the United States, it was "pointed out [to Ellis] that it was against the Japanese culture to publicize something like the response rate, even if it were available" (Ellis 2004b). Since doing so would be tantamount to admitting "'the limitation of the (statistical) method' in a government report[,] [it] was unheard of in Japan" (Ellis 2004b). Although Ellis did not pursue this project further on her own, she nonetheless unexpectedly returned to this issue five months later during her prefectural fieldwork.

Specifically, she was soon to discover that all three of her objectives including (1) evaluating the project management style and method of design, (2) learning how survey operations including that of data collection are conducted, and (3) gaining an understanding of non-sampling errors were to some extent intertwined. Challenges inhere not only in the *evaluation* of the data at the SBJ, but also in the data *collection* process at the local level as well. To be clear, Ellis perceives that the power sharing structure itself between both levels of public administration is likely the critical explanation for the challenges to effective data collection at the ground level. As she explains,

> The real reason for difficulty in collecting data on response quality in Japan . . . lies in the fact that enumerators collect questionnaires from every household [and] . . . are asked to check the completeness and accuracy of the responses before they leave the household. If there is any correction on the form by an enumerator, then only the corrected information is entered into electronic data. . . . [Thus] the Statistics Bureau . . . was afraid they might exacerbate their relation[s] with the local governments by increasing clerical work of the enumerators even further if they were to collect such additional data.

As this reveals, intragovernmental relations were apparently given priority above the requisites of sound scientific data collection in this case. In fact, Ellis continued to observe difficulties of coordinating the local data collection process with the centralized administration of the SBJ. This was not a challenge Ellis had experienced during her career with the U.S. Census Bureau.

To further illustrate this point, Ellis (2004b) provides an extensive comparison between census data collection in the United States and Japan, which is excerpted below. First she explains that local government administrators are able to exert far more power in the data collection process in Japan's national surveys than in the United States,

> It is very clear from all the trips to local governments on a regular basis and all the meetings held at the headquarters for local officials every few months that a lot more input from local governments goes into the planning of the census in Japan than in the U.S. This is because the Statistics Bureau depends completely on the local governments to collect data for all their surveys while the U.S. Census Bureau in general has a lot less dependency on local governments for data. . . . For the U.S. Decennial Census, for example, local governments are involved in the census mainly through pre- and post-evaluation of the address lists in their areas. They also become active in attracting enumerators from their own communities, either for the non-response follow-up or for the post-enumeration survey. Most census forms are mailed directly to each household and they are mailed back to the Central Processing Office in Jeffersonville, Indiana so no local governmental officials are involved in data check[ing].

The consequences for increasing the power of local government administrators in the data collection process, however, is significant. As Ellis clarifies,

> One of the drawbacks of the local governments' involvement [in Japan] is that it slows down the decision-making process considerably. Trying to get consensus not only within the Statistics Bureau but also from cities and prefectures around the country is, to be sure, time-consuming. For example, it has taken more than 6 months for the Statistics Bureau to decide what kind of envelope to use to put the census form that will be hand-delivered by enumerators. In [the] old days (1995 and earlier), no envelopes were used because enumerators would pick up the census forms at every household and check, in front of the respondents, if all questions were correctly answered. Now, because of privacy concerns, a lot of respondents do not want the enumerators to check the forms. Instead, they want the forms to be forwarded to prefectural offices and checked there. The Statistics Bureau however wants to discourage respondents to put a census form in an envelope because this will prevent an enumerator to check the form, thereby increasing the adjudication workload by prefectural employees. So, they have debated, among others, whether to use a seal instead of an envelope, whether to give out an envelope only if a respondent asks [for

> one], and if they are to use an envelope, which envelope guarantees non-transparency to prevent enumerators from peeking in (Ellis 2004b).

As this reveals, the laborious process to achieve consensus between the local and central government administrators has become a driving force behind policymaking rather than a purely data driven outcome. Likewise, this process has become hedged with significant procedural hurdles which slowed solution driven policy implementation.

Once again comparing the United States and Japan, Ellis (2004b) explains how the same problem might have been approached in the United States,

> What the [U.S.] Census Bureau would have done in this situation is, they would have conducted statistical experiments in test censuses to [ascertain] which option would result in the best response rate with the best accuracy and then make a final decision on which option to adopt.

Hence, test cases in the U.S. policymaking context to evaluate how to produce the highest response rate would replace the tedious process of consensus building between central and local government administrators in Japan. As she explained, the SBJ

> would have [representatives from local governments, . . . the [forty seven] prefectures in Japan, . . . come to Tokyo once every month . . . to get the consensus among the local staff before they could do anything or make a decision about which envelope to use. And anything that they decide in data collection, they have to have the O.K. from the local governments. It's so difficult (Ellis 2009, recorded telephone interview).

Thus, Ellis notes that finding a workable way to incorporate the perspectives of local municipalities into the policymaking arena of the central ministry was challenging (Ellis 2009, recorded telephone interview). Within the SBJ, major policymaking decisions continued to be handed down hierarchically from the senior administrators in the SBJ who were also those with the greatest education and training having passed the Level 1 civil service examination. Nevertheless, with regard to clerical functionality, Level 2 and 3 employees weighed in on a daily basis. With regard to the SBJ's larger policy network, representatives of local governments also played a well-defined role in policy implementation and data collection rather than at the formulation stage as their willing cooperation was necessary to achieve desired policy outcomes.

As such, this case study once again confirms the expectations of the research question in that the absence of multiple principals decreases opportunities for bureaucratic drift in this case. The effect is significantly reduced bureaucratic discretion to get the job done. Specifically, this policy context is clearly noted by the absence of multiple principals and instead the presence of multiple agents. In their absence, all forty-seven prefectures along with the central government exert policy pressure. Consequently, rather than enhanced bureaucratic discretion, policy gridlock results over relatively minor items such as the proper forms to use consumes the involved agents. As a result, unlike in instances of enhanced bureaucratic drift, the agents in this case lacked the authority to pursue their preferred policy outcomes.

CASE STUDY 1: SUMMARY

The participant-observer experiences of these two Mansfield Fellowship alumni at the Statistics Bureau and Telecommunications Bureau of the Ministry of Internal Affairs and Communications respectively demonstrate strong evidence to support the research question. Starting with the Telecommunications Bureau, the presence of multiple principals increased bureaucratic drift causing the discretionary authority of bureaucrats to rise. In the second case of the Statistics Bureau, the paucity of principals, but plethora of agents decreased bureaucratic drift causing the discretionary authority of bureaucrats to fall. The next chapter turns to the second case study of the Ministry of Foreign Affairs (MOFA).

NOTES

1. See MIC's website at http://www.soumu.go.jp/english/soumu/index.html (accessed June 28, 2011).

2. See MIC's website at http://www.soumu.go.jp/english/index.html (accessed June 28, 2011).

3. See MIC's website at http://www.stat.go.jp/english/info/guide/2009ver/04.htm#01 (accessed August 30, 2017).

4. The SBJ also calls upon the expertise of academicians through an external body at that time called the "Statistics Council." Interestingly, the Statistics Council immediately perceived the value from a validity and scientific standpoint of collecting this additional information for the 2000 National Census. Although a meeting ensued between the Statistics Council and the SBJ, Ellis observed that a resolution of the problem remained elusive (Ellis 2004b). Subsequently, the "Statistics Council" was replaced with the "Statistics Commission" in October of 2007. As the SBJ clarifies, "The national statistical systems are constantly evolving all over the world, and several countries recently enacted or renewed the legislations giving foundations for official statistics. Japan is no exception, and a new act was passed by the Parliament in May 2007 to replace the Statistics Act that had been in effect since 1947. By the new act, 'Statistics Commission,' a high-level consultative organ that assumes a leadership for development of official statistics, established in the Cabinet Office in October 2007, and after preparatory work for over a year the new act [was] fully enacted in the spring of 2009" (See: http://www.stat.go.jp/english/info/meetings/sympo/sympo.htm, accessed August 30, 2017). In operation, Ellis explains that "one of the roles of the Statistics Commission is to oversee the government effort to overhaul the Japanese statistical system as a professional, neutral, and impartial third-party organization, consisting of thirteen experts. [The] SBJ has a monthly meeting with the Statistics Commission to go over their work . . . [and] occasionally invites academic experts to [present] . . . on specific topic[s] of interest" (Ellis 2011).

5. In marked contrast, to obtain employment in the U.S. Census Bureau, the job description for Ellis's position required her to have a degree in mathematics or statistics with a certain amount of commensurate coursework. In Japan, employees may have diverse educational backgrounds and she learned that a number of those at MIC completed legal programs at the university level.

6. Notably, Ellis was a helpful asset for the SBJ in this regard. For instance, the SBJ was delighted to learn from Ellis requested information about the U.S. Census Bureau's Internet site, "American Fact Finder." After returning to the U.S. Census Bureau in October of 2004, Ellis has continued to facilitate the exchange of information between the SBJ and the U.S. Census Bureau and expedited fact-finding missions of Japanese government personnel to learn more about U.S. approaches and processes firsthand.

7. One example of how change in one part of the system might require changes in another part of the system in a different context was the 2002 movement to consolidate and streamline various economic surveys that were conducted by different government agencies at different cycles into an economic census. The ultimate goal was to produce more timely and higher quality data that were essential for the production of national accounts. There were varying degrees of enthusiasm and resistance for the big change among the agencies involved in the beginning. It took numerous meetings and lengthy efforts for every party to come to an agreement on the content and the timing of the economic census. The second-phase of the economic census, "Business Activities Survey," will finally be carried out in February 2012, after having been pushed back from the initially scheduled 2007 (Ellis 2011; Ellis 2009, personal communication).

Chapter Three

Case Study 2

Ministry of Foreign Affairs (MOFA)

INTRODUCTION

This chapter evaluates the fundamental issue of bureaucratic drift to the Japanese democracy by reviewing the Ministry of Foreign Affairs (MOFA). Specifically, this chapter begins by sketching MOFA's relationship with political principals over the past few decades drawing upon Robert Scalapino's (1977) edited volume along with George Ehrhardt's (2009) study applying principal agent theory to MOFA. Ehrhardt (2009) analyzed foreign policymaking in Japan from 2001–2003 using MOFA as his case study. His research affirms that the presence of multiple principals (independent variable) in foreign policymaking increased opportunities for bureaucratic drift (dependent variable) during that timeframe. After concluding this review, this chapter introduces two additional case studies involving MOFA which do not implicate multiple principals. In these two cases, Mansfield Fellows were posted at MOFA including Michael Panzera and Jeffrey Seay. As predicted, both cases affirm that the dearth of multiple principals decreased opportunities for bureaucratic drift. The attendant causal effect has decreased bureaucratic discretion to problem solve in significant policy affairs related to domestic labor shortages and nuclear proliferation.

APPLYING PRINCIPAL AGENT THEORY TO MOFA

If there is any Japanese ministry one might expect to be involved in a controversial policy contexts implicating issues salient to the general public, it is the MOFA. Unsurprisingly, MOFA has many different bureaus that handle a wide array of issues such as the Consular Affairs Bureau, Economic Affairs Bureau, European Affairs Bureau, Foreign Service Training Institute, Foreign Policy Bureau, Intelligence and Analysis Service, International Cooperation Affairs Bureau, International Legal Affairs Bureau, Middle Eastern and African Affairs Bureau, and Public Diplomacy Department. [1]

Additionally, MOFA engages in various public diplomatic projects. For one, MOFA encourages intergovernmental relations such as facilitating cooperation with United Nations entities via the Japan Foundation.[2] For another, MOFA is involved with sundry international cooperative efforts from the Asia Cooperative Dialogue through periodic summits of foreign ministers to the 18th Asia Pacific Economic Leadership Meeting held in Yokohama in 2010 (APEC 2010). Moreover, MOFA advocates exchange programs such as the widely known Japan Teaching and Exchange Program, more commonly called the "JET" Program. This program allows foreigners to teach English while living in Japan and is heavily promoted in American universities. Finally, MOFA is also involved in many other visible activities such as bidding for the 2016 Olympic Games to be held in Tokyo along with highly controversial immigration questions that will be discussed below.[3]

As this may suggest, MOFA's primary mission is public diplomacy seeking to enhance Japan's image abroad. As MOFA explains:

> In order to conduct foreign policy smoothly and effectively, it is essential to promote an understanding of Japan among the general public overseas and to enhance their image of and sense of affinity toward Japan, in addition to appealing against the backdrop of . . . democratization around the world, public opinion has a growing influence on foreign policy. Accordingly, many observers have stressed the importance of "public diplomacy"—direct appeals to citizens and public opinion in other countries in cooperation with the private sector.[4]

The most remarkable aspect of this broad mission statement is its scope as it lends the impression that MOFA actually unilaterally conducts foreign affairs in Japan. However, Article 73 of the Constitution of Japan vests the power in the Cabinet Office to "[m]anage foreign affairs."[5] While administrators have certainly exercised critical roles in policymaking, analysts have noted MOFA's relative weakness to conduct foreign policy vis-à-vis elected officials even as far back as the 1970s. As Haruhiro Fukui (1977, 4) explains,

> There are factors working against bureaucratic control of policy-making processes, especially in the foreign policy arena. . . . Among the bureaucrats compartmentalized into ministries and agencies, those in the Foreign Ministry are relatively weaker, if not the weakest, in terms of political power and clout.

He goes on to note that MOFA was disadvantaged both in terms of funding and staffing needs ranking as the second smallest (Fukui 1977, 4). Likewise, political principals have been known to ignore the advice of administrators choosing instead to pursue their own preferences (Fukui 1977, 4). Finally, Fukui identifies significant inter-ministerial disputes that undermined MOFA's authority as well (1977, 4). Importantly, Fukui (1977) explains that MOFA's authority at that time generally waxed and waned with the saliency of the issue. Whereas more controversial issues enhance principals' involvement, more routine issues decrease principals' involvement. Even so, "[i]n either case, the number of participating officials is not very large" vis-à-vis other ministries such as the Ministry of Finance (MOF) in other policy contexts (Fukui 1977, 8–9).

This trend in favor of the power of politicians continued in administrative forms over the 1980s and 1990s (Ehrhardt 2009, 627). However, the seeds had been sown in the early 1990s for a rebalancing of power with decreased involvement of multiple principals. Specifically, the 1990s also saw the appearance of enhanced interparty competition throughout East Asia (Ehrhardt 2009, 625) (see chapter 1 *supra*). As Ehrhardt (2009) explains, with this interparty competition came significantly enhanced conflict amongst politicians and was particularly evident with regard to MOFA.

Ehrhardt (2009, 635) specifically analyzes this principal agent problem in the context of Foreign Minister Tanaka Makiko's attempt to rule MOFA with an iron fist. As the daughter of the now infamous Tanaka Kakuei, she was determined to bring her father's resolution—conceived in the 1970s—of the Northern island territorial dispute with Russia to fruition (Ehrhardt 2009, 635). However, as the time approached for her assumption of office, MOFA personnel felt that her father's solution no longer matched with foreign policy realities as Russia did not support the plan (Ehrhardt 2009, 635). Thus, just days prior to her arrival as Foreign Minister, a key staffer of the Russia Desk, Jiro Kodera, who was known for his support of the Tanaka plan, was sent off to the London embassy in a bid by MOFA to derail her efforts before she even began by depriving her of a key ally (Ehrhardt 2009, 635). Without going into the full details of the ensuing events well-described by Ehrhardt (2009), the following sketches the very public row between Tanaka and her ministry which resulted.

At first, it appeared that Tanaka would prevail as she reinstated Jiro bringing him back from London and "freezing all personnel changes at the ministry" (Ehrhardt 2009, 635–36). However, unbeknownst to her, MOFA was engaged in embezzlement schemes and by taking the conflict public, MOFA's internal operations which had enjoyed a clean image were suddenly under close scrutiny (Ehrhardt 2009, 635–36). As a result, in March 2001, a MOFA logistics official, Matsuo Katsutoshi, was arrested on suspicion of embezzlement of secret government funds (*JJTI* 2001). A year later in April of 2002, MOFA was rocked by a significant scandal implicating thirty-four officials. As an article at the time announced, "Foreign Ministry Bureaucrats Have Much to be Ashamed of" (*Asahi Shimbun* 2002b). Consequently, Tanaka and MOFA became quickly embroiled in an ever widening controversy. In fact, an all-out war between Tanaka and MOFA's top brass ensued as "ministry officials took the battle to the media" alleging that Tanaka was incompetent and ill-suited for the job (Ehrhardt 2009, 636). The situation continued to spiral out of control into a public relations nightmare as Tanaka launched counterstrikes of her own (Ehrhardt 2009, 636). For instance, *Asahi Shimbun* (2002b) (i.e., the widely read *Morning Sun Newspaper*), adopted a harsh in its tone insisting that MOFA personnel were ill informed about the domestic issues facing Japan at the time having grown "elitist" believing "that they alone truly understand diplomacy."

Taking a short step back, the broader media context in which MOFA operates bears note. Interestingly, while MOFA's activities are significant for maintaining Japan's international standing, they do not necessarily impact the daily lives of Japanese citizens in a directly observable way. Thus, as a consequence of the attention of the public being spread thin over a wide array of policy topics particularly given its staffing constraints, MOFA's internal operations had essentially avoided public notice up to this point. Examining newspaper articles from the *Asahi Shimbun* over the period of 1997–2010 lends support for this observation. While MOFA is mentioned in many articles involving topics from international security to immigration policies, MOFA *itself* was rarely the object of review over this timeframe until the scandals erupted.[6] Yet, as table 3.1 below reveals, the number of articles about MOFA reached an all-time high in the context of this bitter feud between Tanaka and her ministry.

Eventually, these battles reached such a fever pitch that prime ministerial involvement became imperative. As Ehrhardt (2009, 627) clarifies,

> Beginning in 2001 . . . Prime Ministers [Junichiro] Koizumi and Abe Shinzo intervened in battles between cabinet ministers and their administrative vice-ministers (i.e., the ranking bureaucrat [in the ministries]), undercutting politicians' control over the bureaucracy. This shift is still not integrated into the current literature.

Thus, recognizing that policy gridlock had resulted from allowing multiple principals to exert authority over MOFA, Koizumi decided to "constrain political control over the bureaucracy to prevent intervention by rivals" (Ehrhardt 2009, 626).

In the end, the newly appointed Foreign Minister, Yuriko Kawaguchi, took action against high ranking members of his ministry after it came to light that Lower House legislator, Muneo Suzuki, had influenced policymak-

Table 3.1. Ministry of Foreign Affairs Cited by the *Asahi Shimbun* (1997–2010)

Year	Ministry of Foreign Affairs *Asahi Shimbun* Hits
1997	103
1998	239
1999	298
2000	1
2001	37
2002	348
2003	250
2004	223
2005	137
2006	178
2007	108
2008	107
2009	91
2010	182

Source: Author.

ing in MOFA with regard to the sensitive issues of negotiating with Russia over disputed island territories.[7] Additionally, reforms were proposed to boost the morale of MOFA and set it back on track. As a writer for the *Asahi Shimbun* (2002a) explained in this case,

> One proposal in the reform plan that should be given priority is to fill key positions, such as those of ambassadors to Japan's major trade partners, from the private sector. There is no need to allow professional diplomats to enjoy a monopoly over Japanese diplomacy. The use of human and other resources from the private sector would help invigorate diplomatic activities. That must be clear to Kawaguchi, who once worked for a private business.

Specifically, Ehrhardt (2009, 633) points out that although the late 1990s saw reforms in the name of reducing bureaucratic discretion, "underneath the rhetoric of bureaucratic accountability was a battle between political principals." By increasing the power of Diet member's to oversee administrative affairs through two vice minister posts, requiring testimony of ministers and vice ministers to Diet questioning on a regular basis, merging a number of the ministries, and increasing the power of the prime minister through staff increases and other powers (see chapter 1).

Once again, however, in accordance with the expectations of the research question, the eventual outcome of involving multiple principals increased opportunities for bureaucratic drift (Ehrhardt 2009, 633). As Ehrhardt (2009, 634) explains, "[t]hese changes left the bureaucracy. . . now caught between three political masters, each demanding different things." Ultimately, those interests collided to such a degree, that Prime Minister Koizumi intervened to restore the balance in favor of enhancing MOFA's power to resist the interference of principals. As Ehrhardt (2009, 639) explains,

> In fact, in June 2003 in Diet testimony, [Koizumi] explicitly praised the ideals of bureaucratic excellence in the responsibility model, and claimed that his actions were intended to let bureaucrats display their creativity and discretionary power free from political influence.

This is consistent with the presentation of Koizumi in chapter 1 as a greater advocate of the bureaucracy than most realize today.

With this preliminary grounding in MOFA's policy network along with additional support for the research question, two more case studies are introduced below to bolster these observations. In accordance with the predictions of this project, these case studies clarify that the *absence* of multiple principals (independent variable) decreases opportunities for bureaucratic drift (dependent variable). In turn, the observed effect is that discretionary authority of agents *decreases*. Specifically, the first case study comes in 2008–2009, five years after Koizumi had restored the power of MOFA and the involved ministries to "display . . . discretionary power free from political influence" as noted above. The second case demonstrates the policy arena in 1999–2000 and thus prior to the scandals MOFA faced. The latter example is thus equally important as it transpired near the end of MOFA's long period without significant public scrutiny and thus is notable for the absence of multiple principals due to the neglect of close scrutiny of MOFA's internal affairs at that time.

Michael Panzera, Esq.

Michael Panzera, a trial attorney for the National Courts section of the Civil Division in the U.S. Department of Justice, completed his Mansfield Fellowship (2008–2009) at the MOFA, the Ministry of Economy, Trade, and Industry (METI), along with the Legal Research and Training Institute of the Tokyo High Court under Chief Judge Sonou. He was also an invited participant in the National Personnel Authority's (NPA) administrative training program, studied at the Legal Research and Training Institute for judges and prosecutors in Japan, and served as a Visiting Research Scholar of the Graduate School of Law and Politics at The University of Tokyo. At the time of this writing, he litigates international trade cases in the Court of International

Trade and the Court of Appeals for the Federal Circuit and is an adjunct professor at The George Washington University Law School. Prior to joining the legal profession with his JD from the University of Washington and BA in Mandarin Chinese from Georgetown University, Panzera served as an interpreter and translator in Chile, Washington, DC, and Taiwan. Thus, it is perhaps unsurprising that he also assisted METI in document translation from Japanese to English as a Mansfield Fellow.

Panzera's objectives during his Fellowship "included gaining a firsthand understanding of the Japanese approach to legal analysis, negotiation and conflict resolution with particular emphasis on international trade policy" (Cottingham-Streater 2010, 3). The objectives of his Mansfield Fellowship, like those of Ellis and Miller described in the previous chapter, once again focus on understanding the contours of this particular policy arena.[8] Finally, Panzera's level of Japanese proficiency was sufficient for the task as the Section Chief at MOFA instructed everyone to speak to him only in Japanese during his stay and he was "able to communicate with everyone about the issues that [were] being negotiated" (Panzera 2008).

Economic Partnership Division

To begin, while Panzera was involved in assisting on a variety of foreign affairs related projects during his time at MOFA, one of his early assignments involved assessing the recognition of credentials between states through a mutual agreements program between foreign states to allow qualified foreign professionals to work in Japan on an as-needed basis (Panzera 2011, personal communication). As Panzera details in his Mansfield Progress Report, he worked on,

> Japan's Economic Partnership Agreement (EPA) with Indonesia in which [Japan] agreed to allow Indonesian nurses and caregivers to reside temporarily in Japan under certain restrictions. Specifically, Chapter 12 of the EPA provides that nurses may enter Japan under a "designated activities" visa for a period of no more than 3 years, provided that they are already licensed in Indonesia and have more than 2 years experience. Caregivers may enter Japan for a period of no more than 4 years (Indonesia currently lacks a certification process for caregivers). The EPA further provides that Indonesia will send a total of 1,000 nurses and caregivers to Japan by 2009. Once nurses have been selected for entry, those workers spend the first 6 months in language and social training. (Panzera 2008).

While this may seem to be a relatively straightforward topic, immigrant workers entering Japan is a controversial issue appearing repeatedly in the popular media. In fact, there is some resistance both in the public and in the ministries to the argument that Japan requires additional laborers from outside the country during the present economic downturn. For instance, as

Panzera (2011, e-mail) clarifies, "the Ministry of Health, Labor, and Welfare (MHLW) had expressed strong opposition to certain aspects of this plan, based upon a belief that labor shortages could be met domestically, while the MOFA was under intense pressure to reach an agreement with its foreign counterparts on this issue." Thus, as the outcome of this dispute rested with the two ministries (i.e., MOFA and MHLW), it is clear that this case was to be resolved in the absence of multiple principals.

In the middle of this inter-ministerial policy dispute, Panzera arrived at MOFA and was assigned "to research other countries' approaches to allowing foreign guest worker to immigrate temporarily pursuant to the terms of . . . bilateral agreements, either agreements limited to specific sectors or as part of an overall trade agreement" (Panzera, September 8–30, 2008, 2). However, Panzera initially wondered whether this project might be "busy work" as the full extent of this policy dispute between MOFA and MHLW was not clarified. Thus, after some time, Panzera directly inquired of his department chief why he was tasked with a "services-related issue if I am not in the services section of MOFA, which usually handles this issue" (Panzera 2008).

In response, Panzera was given to understand that MOFA was hoping to use him as an acceptable intermediary between the two ministries since if he were to lead a presentation on the issue in a joint session of the two ministries, his point-of-view would be taken as that of an outsider rather than representing the perspective of the Economic Partnership Division of MOFA. In this way, the Section Chief of his division in MOFA could lend a hand to the Section Chief from the Services division of MOFA thus potentially leading to a larger accord or improved understanding when the Services division met with MHLW. In other words, Panzera was to be a liaison not only between two divisions in MOFA, but also between MOFA and MHLW. Through conversations with his supervisors at MOFA, he was informed that the Economic Partnership Division in which he was placed,

> handles the overall negotiation strategy. . . [Thus,] perhaps the negotiators need to become more familiar with other countries' agreements and innovative ideas used therein in order to exert some pressure on the Services section, which might not be thinking "outside the box" as much as it could to devise options for negotiation (Panzera 2008).

Panzera was first tasked with preparing a presentation in Japanese for representatives of both divisions of MOFA. Interestingly, however, Panzera was very reluctant to provide a direct recommendation on the matter as "it might be presumptuous . . . to say what would be most suitable to Japan's economy." Thus, Panzera structured his talk with the help of his section chief to speak in more general terms rather than providing a direct recommendation

(Panzera 2008, 1). Yet, much to his surprise, upon delivering his presentation in November, this was the first question the Services Chief posed. As Panzera explains,

> The ironic thing was that I had been trying to avoid making a recommendation in my speech, but the Services Section Chief's first question was which approach did I think was the most effective. After giving a disclaimer that I am not an expert in the field, I explained that, in my view, there was no one perfect approach and that each approach had to be adapted for particular characteristics of the countries, but that some options seemed more effective than others in terms of creating tangible, verifiable results, and were politically more palatable because they incorporated domestic sectors in the decision-making process (for recognizing other countries' credentials, etc.). I also explained that the flexibility of the regional "occupations under pressure" lists helped assuage domestic concerns because as certain professions could be filled with domestic labor, those professions could be taken off the list (Panzera 2008).

Panzera discovered that not only was his speech well-received, but that he would be presenting on the same topic of labor shortages and foreign guest workers to the World Trade Organization (WTO) section of MOFA and subsequently to the MHLW.

> It was explained to me that the MHLW does not fully accept that there is a need to recruit workers from abroad, so this presentation might help them think more creatively about filling labor shortages. The section chief emphasized that I should not hold back in giving analysis on comparing the relative merits of the program . . . (Panzera 2008).

The next month, Panzera did as requested by presenting at a joint ministerial conference on various themes with regard to alternative international aspects and strategies that might be employed to solve the problem of worker shortages. As he explains at length,

> About 40 people from several ministries attended the speech, including the Ministry of Justice [MOJ] and [the Ministry of Economy, Trade, and Industry]. As requested, I covered the basics of foreign strategies and then focused on general themes. . . . After the speech, [we had] about 20 minutes for Q&A. Several people had interesting questions about exploitation of workers, and regional programs. However, no one expressed an opinion about the strategies and the MHLW representative mostly focused on why the programs could never work in Japan, rather than brainstorming about how they could be adjusted to adapt to Japan. For example, they were adamantly opposed to any regional variation in recruitment schemes, such as using regional lists of occupations under pressure to allow for variegated levels of relaxed recognition requirements. However, the only reason they gave for their opposition was that

> it was centrally administered, which is unconvincing because the other countries employing regional lists are also centrally administered (Panzera 2008).

Clearly, the reception of his presentation was quite negative by MHLW. Even so, in post presentation discussions with the representative from the Services Section, MOFA remained hopeful that "the speech was very useful in getting MLHW to consider different approaches" (Panzera 2008). Indeed, from a substantive standpoint, MOFA perceived that some of Panzera's recommendations were even "convincing" including "the theme of incorporating the private sector into the decision-making process as a means of creating an effective recognition program, but also forestalling dissent once the program goes into effect" (Panzera 2008).

Panzera further reflects on the underlying complexities in this policymaking arena with the agents of both ministries lacking the requisite discretionary authority to decide the dispute alone. First, as Panzera clarifies, MOFA in this case was a relatively unified actor. His initial presentation before the Chief of the Services Division of MOFA was feasible as between different divisions in MOFA due to the cooperative spirit within the ministry (Panzera 2008). At this stage, Panzera's intervention was welcomed by the corresponding chief in the Services division.

At the same time, as between MOFA and MHLW, the opportunities for bureaucratic drift narrowed given the absence of multiple principals in this policy context. Instead, as predicted, in the absence of multiple principals, the authorized agents were enmeshed in policy gridlock with both sides lacking the discretionary authority to decide how to make use of foreign laborers to alleviate domestic labor shortages. As such, MOFA concluded that "that the only way of breaking the impasse was to show [MHLW] the outsider's perspective and different ways that this has been approached" (Panzera 2008, personal communication). The fact that MOFA hoped to use Panzera's presence to jumpstart inter-ministerial dialogue is telling. As Panzera (2011, personal communication) opines in his final assessment of the resulting policy gridlock,

> bureaucratic sclerosis just doesn't lend itself to negotiation. That's why you're seeing these trade agreements are so limited in scope. It's not that they don't want to have a wide range scope. . . . I think the reason is because those inter-ministerial disagreements just can't get easily resolved.

Thus, decreasing opportunities for bureaucratic drift in the absence of multiple principals ultimately was manifested in significantly constrained bureaucratic discretion to resolve policy disputes.

Jeffrey Seay

Perhaps one of the most interesting aspects of Panzera's Fellowship at MOFA is that it is not without precedent. Eight years earlier in 1999–2000, Jeffrey Seay also completed his Mansfield Fellowship experience at MOFA. Just like Panzera, Seay was called upon to serve as a liaison by MOFA to ease intergovernmental conflict. Now a noted novelist, Seay is the former Head of the Security and Fraud Risk Department at HSBC Japan where he was responsible, *inter alia,* for planning, development, implementation, and management of physical and information security risk, IT security, business security, and fraud risk and investigation. He is the former head of Corporate Security and Business Continuity for Japan and Korea at Deutsche Bank. Prior to that position, he served for just over twenty years as a special agent at the Naval Criminal Investigative Service (NCIS). He earned his law degree from Whittier Law School. As a Mansfield Fellow (October 1998–September 2000), he completed his stint at the MOFA, (then) Ministry of International Trade and Industry (MITI), National Police Agency, and the Tokyo Metropolitan Police. NCIS approved of Seay's participation in the MFP as "an opportunity for the expansion of its contact base and to promote recognition . . . of the NCIS mission by the Japanese government" (Seay 2000, 1).

Arms Control and Disarmament Division

Upon arriving at MOFA, Seay discovered that his initially intended projects would not be possible to accomplish due to unforeseen security concerns. His first choice placement in the Arms Control and Scientific Affairs Bureau had been rejected by MOFA due to the "sensitive nature of the work conducted there" (Seay 2000, 3). This was so despite the non-disclosure agreement that Seay had completed and the MFP safeguards in place (see chapter 1). Consequently, his Fellowship took a new direction under the Arms Control and Disarmament Division. As Seay (2000, 2) explains,

> The Ministry of Foreign Affairs had helped establish a non-governmental organization called the Center for the Promotion of Disarmament and Non-Proliferation (Center) within the Japan Institute for International Affairs three years prior. Apparently, things had not been going well at the Center and the matter had created a rather delicate political situation between the Center and our Division. What those issues were he did not make clear at the time but [his supervisor] wanted to do something that would help promote the Center. He indicated that funding had been made available through the Division for a one-day international symposium to be "sponsored" by the Center. [Seay's supervisor] felt that I was just the person to get the ball rolling.

The task immediately proved more challenging than Seay had anticipated for although the planned event was to promote the Center, he was essentially on his own in the process as the nongovernmental organization (NGO) did not provide him any support in the task of finding appropriate speakers including a keynote speaker, moderator, and five international experts. In addition, he was tasked with including topics that would "focus on Asian Affairs but would also include issues involving Russia and the new independent states" (Seay 2000, 4). Moreover, Seay was expected to prepare a formal proposal for the event. Interestingly, many of his colleagues in MOFA were doubtful whether he would be able to accomplish the project considering the complex negotiations that such an undertaking would require. As a result, unlike the strong support Panzera received, Seay generally worked alone on the assignment for several weeks.

In the meantime, he was called repeatedly off task to assist with document translation as well as suddenly having to readjust the focus of the symposium when the U.S. Senate failed to ratify the Comprehensive Test Ban Treaty (Seay 2000, 5). However, Seay was most flexible and amenable to this theme change as he did not want the event to compete with a Japan Defense Agency conference in October of 1999.

By December, Seay had made significant progress and had secured sufficiently credible speakers including Michael Krepon from the Henry L. Stimson Center in Washington, DC. Consequently, MOFA and the Center unexpectedly had to provide Seay with significant logistical support to appropriately launch the conference. Hence, a team was assigned to assist in the project and the Center also became an active participant. The event garnered the attention of the news media including the *Yomiuri* newspaper and NHK television on the 9:00 pm evening news (*The Daily Yomiuri* 2000a; *The Daily Yomiuri* 2000b). In the end, the nonproliferation symposium on February 24, 2000, was an astounding success pulling in over 400 attendees including high ranking officials and representatives from 35 countries. The Minister of Foreign Affairs, Mr. Kono, likewise held a special reception afterwards for 70 distinguished invitees including Japanese Ambassadors and foreign diplomats (Seay 2000, 8).

Seay eventually determined that the reason for the Center's lack of strong support and involvement was due to its funding stream. In short, since MOFA was picking up the Center's tab because the Center employed many retired administrators, the Center was not responsible for self-generating funds through publications, endowments, and other fundraisers (Seay 2000, 8). Hence, there was little incentive to produce results. Thus, Seay's involvement helped MOFA to maintain a positive public persona and to be at the forefront of an important foreign affairs topic while simultaneously assisting in reemploying retired administrators and helping all those involved to save face.

While there are some differences, the similarities between Panzera and Seay's experiences are striking. As for the differences, unlike Panzera's lack of experience in immigration affairs, Seay's expertise in matters related to nuclear nonproliferation certainly aided in his identification of experts and in contacting the relevant panelists and the keynote speaker along with constructing the appropriate topics of discussion for the symposium and preparing the proposals for MOFA. In other words, while the agency officials apparently lacked both the will and discretionary authority to build a coalition of participants together in this policy setting, Seay's independent experience and connections proved sufficient to break the gridlock in this instance. Yet, the similarities between the two experiences are even more striking. Once again, as predicted, due to the absence of involvement by political principals, opportunities for bureaucratic drift declined. This seemed to preclude progress by the involved policy actors in facilitating dialogue and resolving a brewing public relations problem. As such, only by tasking an outsider—albeit halfheartedly—with the responsibility of arranging a joint conference was any resolution through dialogue possible.

CASE STUDY 2: SUMMARY

Although serving at divisions in MOFA eight years apart, the similarities between the two Mansfield Fellowship case studies is remarkable. In both instances, bureaucratic discretion at MOFA was reduced due to the absence of political principals. In the first instance, Panzera was placed at MOFA in 2008–2009 following the return of additional authority to the ministries by Koizumi four years earlier. In the second instance, Seay was placed at MOFA in 1999–2000 near the end of an extended period of MOFA's high reputation due to neglect from outside scrutiny.

As predicted, in the absence of multiple principals with regard to the specific issues facing their respective divisions in MOFA, the divisions in both cases experienced policy gridlock. Thus, opportunities for bureaucratic drift declined and the causal effect was decreased administrative discretion to resolve disputes. In both cases, MOFA sought to involve Panzera and Seay respectively in enhancing dialogue between the ministries and agents involved. This was apparently due to the inability of the Japanese agencies involved to achieve progress alone in encouraging dialogue on issues of significant national importance including resolving domestic labor shortages and nuclear proliferation respectively.

Equally noteworthy was that in both cases, Panzera and Seay employed similar approaches to counteract bureaucratic drift: namely, seeking the enhanced involvement of outside stakeholders. In the first case, MHLW and MOFA were advised to end gridlock over visas for foreign workers by including corporate interests. In the second case, enhancing MOFA's leadership on nuclear proliferation was made possible by involving the media in publicizing the joint conference. In both cases, the absence of multiple principals led to policy gridlock and thus decreased opportunities for bureaucratic drift. As a result, agents experienced *decreased* discretion to problem solve. Hence, this case study affirms that in the *absence* of multiple principals, opportunities for bureaucratic drift decline.

NOTES

1. See: http://www.mofa.go.jp/about/hq/org.html (accessed August 30, 2017).
2. Ministry of Foreign Affairs. "Public Diplomacy/Culture." See: http://www.mofa.go.jp/policy/culture/index.html (accessed August 30, 2017).
3. MOFA. "Tokyo 2016—The Olympic Games. Tokyo's Bid for Hosting 2016 Olympic and Paralympic Games."
4. Ministry of Foreign Affairs. "Public Diplomacy/Culture" See: http://www.mofa.go.jp/policy/culture/index.html (accessed August 30, 2017).
5. See http://www.kantei.go.jp/foreign/constitution_and_government_of_japan/constitution_e.html (accessed August 30, 2017).
6. Reviewing the hundreds of citations MOFA received leading up to the national administrative reforms of 2000, indicates a majority referenced MOFA's international activities After the 2000 administrative reorganization went into effect, the number of articles mentioning MOFA sharply declined for two years. It only shot upwards to the highest mark ever in 2002 at 348 with the scandals. Yet, it is important to note that the remaining spike in articles in 2002 was due to the war in Afghanistan. Notably, this timeframe also coincided with the global war on terrorism and thus this was a singularly unfortunate time for MOFA to be embroiled in controversy given its larger policy responsibilities. Indeed, during 2002, MOFA was involved in numerous security related matters. For instance, MOFA was particularly focused on Japan's role in Afghanistan and domestic terrorism threats. A careful review of the articles from the *Asahi Shimbun* (i.e., *Morning Sun Newspaper*) (see table 3.1), for example, revealed that over 100 of the articles in 2002 are attributable to themes involving terrorism, security, nuclear weapons, the Middle East, and Afghanistan.
7. In this regard, MOFA appears to be a fairly typical foreign affairs ministry vis-à-vis other established democracies. In fact, foreign policy analysts might find the above rendition particularly unsurprising as the U.S. State Department has been heavily criticized for elitism, excessive caution, and *clientitis* as its long-time appellation "Foggy Bottom" aptly reveals (Hook 2011, 181–82).
8. This focus is not by accident. Rather, the Fellows are encouraged to make sufficiently broad proposals focusing on decision-making in Japan to permit flexibility of goals and observational methodologies to permit the greatest breadth of experience (Cottingham-Streater 2011, personal communication).

Chapter Four

Case Study 3

Ministry of Land, Infrastructure, and Transport (MLIT)

This third case study evaluates the Ministry of Land, Infrastructure, and Transport (MLIT) by reviewing the Mansfield Fellowships of Anthony Waller and Naveen Rao. As in the case studies in the two preceding chapters, this study begins with a brief history of MLIT, before presenting the experiences of the two Mansfield Fellows. This section finds that due to a merger between ministries with very different approaches to policy implementation, the resultant ministry is internally divided resulting in severe bureaucratic drift for MLIT as a whole. Thus, the research participants selected for this project permit an evaluation of both the former Ministry of Construction (MOC) and Ministry of Transportation's (MOT's) approach to policymaking respectively.

In the first case of the Government Buildings Department, Architectural and Building Engineering Division, which would have fallen under the former MOC, this is a highly technical atmosphere requiring significant evaluation of scientific data and policy expertise to make structural changes and enhancements to government facilities. This division is notable for the presence of multiple principals represented by various societal stakeholders who intervene at various stages in the policymaking process. As such, this project predicts *increased* opportunities for bureaucratic drift with the corresponding effect of enhanced discretionary authority of agents. As the discussion below clarifies, this is indeed the case as manifested through an extraordinarily lengthy, tradition bound, multi-iterative policymaking process that the Government Buildings Department oversees when evaluating changes to government building designs.

The second case study of the Japan Civil Aeronautics Board and then the Aviation Industries Division of the Ministry of Land, Infrastructure, Transport, and Tourism (MLITT) provides an interesting comparison. This division would have fallen under the original MOT rather than the MOC and thus manifests a very different policy milieu. Specifically, although this policy context also involves a multiplicity of principals at the policymaking stage, aviation policy is steeped in controversy rather than consensus. This is due to the strong interest of the Japanese public in the hotly contested and beleaguered airline industry in Japan. Even so, this project once again predicts *increased* opportunities for bureaucratic drift in this policymaking arena. Indeed, when it comes to opening new airports or determining the frequency of flight routes, multiple principals are often vocally involved as they press for their particular interests. The result has been significant bureaucratic drift as agents have enhanced opportunities to pursue their own preferences. Once again, the effect has been enhanced bureaucratic discretion with often irritated principals. As in the two previous chapters, the following first introduces MLIT before discussing the two selected Mansfield Fellows for in depth analysis. Finally, findings of the two studies are compared and contrasted.

INTRODUCTION

The Ministry of Land, Infrastructure, and Transport (MLIT) which includes "Tourism" (MLITT) in its title as of 2010, was created by merging the Ministry of Transportation (MOT), the Ministry of Construction (MOC), and two agencies. The rationale behind the merger was to reform these ministries and agencies, which were perceived to suffer from significant "sectionalism." As the new ministry's first White Paper in 2001 specifically identifies as a "result of integration,"

> By removing bureaucratic sectionalism in the four former ministries and agencies and by exhibiting comprehensiveness including coalition/adjustment with other government organizations, integrated/coalition measures have been promoted in order to develop high-quality measures usefully and quickly (MLIT 2001, 4).

The logic for this reform was driven in large part by soliciting public "[o]pinions concerning MLIT as a whole" (MLIT 2001, 3). Public surveys revealed that the largest number of respondents (30.9 percent) felt "[i]t [was] necessary to remove bureaucratic sectionalism [and thereby] to show the merit of integration" (MLIT 2001, 3). The second largest group of respondents (30.6 percent) held that "[a]dministration [through] . . . integration should be promoted" (MLIT 2001, 3).

Despite endeavoring to decrease sectionalism through merger, however, MLIT continued to be a tale of two ministries. As *The Daily Yomiuri (Tokyo)* newspaper found in 2002, the conflicting policy approaches of the two ministries was hindering effective policy implementation. Calling it by an awkward combined name rather than its new official title, media sources pointed out that

> The Construction and Transport Ministry remains troubled by in-house conflicts between officials from the two former ministries involved in construction and maintenance of infrastructure—the Construction Ministry and the Transport Ministry. The reform of special public corporations undertaken last year included four road-related public corporations under the jurisdiction of the former Construction Ministry, including the Japan Highway Public Corporation. However, Construction and Transport Ministry officials from the former Transport Ministry reportedly remained indifferent throughout, simply following events as bystanders without attempting to work together. As a result, the Construction and Transport Ministry is said to have failed to form a united front to protect its interests (*The Daily Yomiuri (Tokyo)* 2002).

Although this article dates back to 2002, the continuing sectionalism of MLIT's two former ministries continued and became such a significant issue that in 2008, MLIT inaugurated the Japan Tourism Agency and expanded the ministry's official title to the MLITT. The remarkable purpose of this new agency—to help overcome ongoing internal problems—was noted at the time of its commissioning. As *The Daily Yomiuri (Tokyo)* (2008b) newspaper reports,

> The agency—the first government ministry-affiliated body to be established since the Financial Services Agency was established in 2000—was created by upgrading the tourism division of the Construction and Transport Ministry, a division formerly headed by bureau chief-level officials. It is hoped the agency will break down bureaucratic sectionalism and attract more foreign visitors.

In other words, this continuing divide between these two former ministries had grown so significant as to spur MLIT to undergo yet another significant structural and name change. Again, it is also notable that the popular media continued to resist using the official name of the new ministry. In fact, it was initially challenging to discern MLIT's presence in the popular news media over the 2000–2008 timeframe since the media rarely mentioned the new name of the ministry instead referencing either "The Ministry of Construction," "Ministry of Transport," or "The Ministry of Construction and Transport" (e.g., see *The Daily Yomiuri* 2001; *The Daily Yomiuri* 2004; Sato 2007). MLITT today clearly remains internally fractured. The next sections evaluate the Mansfield Fellowships of Anthony Waller and Naveen C. Rao. Once again, Waller and Rao completed fellowships as described below in divisions which would have fallen under the Ministry of Construction and Ministry of Transportation respectively prior to MLIT's merger in 2000.

ANTHONY J. WALLER

Anthony Waller completed his Mansfield Fellowship (2008–2009) in the Government Buildings Department, Architectural and Building Engineering Division of Japan's MLITT after having served as a professional in U.S. federal design and construction for twenty-eight years (Waller 2009). Waller was the Tenant Representative for the renovation and modernization of the Government Services Administration's (GSA's) historic Washington, DC, headquarters. Waller has had a distinguished career with GSA previously serving as the Chief Administrative Officer for the Office of the Chief Architect (OCA) of the GSA which is colloquially known as the "Government's Landlord" (Waller 2011). As a former head of administration, he oversaw a $29.5 million dollar budget for the OCA's Centers/Programs which focused on design, construction, preservation, and aesthetic improvements to U.S. federal buildings. He also served as the National Program Manager for Accessibility working as GSA's Liaison to the U.S. Architectural and Transportation Barriers Compliance Board (a.k.a. "Access Board") which ensures design compliance with the *Americans With Disabilities Act* and the *Architectural Barriers Act*. Among many other positions, he served on the Board of Directors of the International Interior Design Association, the Council of Federal Interior Designers, and the National Council for Interior Design Qualifications, not to mention holding the Chair of the Air Force Interior Design Awards (Worldwide). He is also the past president of the International Interior Design Association Foundation. Finally, he has served as an interior designer and project manager for several federal agencies including the Department of Justice and the Office of the Secretary of the Army, Pentagon.[1]

The U.S. GSA had hosted representatives from MLIT since the 1980s, but MLIT had been unable to repay the kindness to a visitor from the GSA due to U.S. federal regulatory constraints on international travel for GSA officers. However, in an exceedingly rare turn of events due to changes in U.S. presidential administrations, Waller was given permission by the GSA to apply in 2008 to the Mansfield Fellowship Program (MFP). Upon acceptance to the MFP, Waller completed his Japanese language training along with the other Mansfield Fellows at the George P. Shultz National Foreign Affairs Training Center, the diplomatic training institute of the U.S. Department of State in Arlington, Virginia.

Government Buildings Department

The significance of the foregoing background is threefold. First, due to the deep contacts GSA had cultivated with MLIT over the years, unlike typical Mansfield Fellows who completed stints at various ministries and agencies, Waller was the guest of MLIT alone during his *entire* Mansfield Fellowship meaning that he did not travel to alternate ministries or agencies. Second, MLIT also ensured that Waller was provided access to some of the most secure and exclusive venues from examining the design of Prime Minister Aso's office in parliament to the temperature controls for the Japanese Emperor's bonsai collection. He was also taught the layering process of gold leaf on the doorknobs during his visit to the famed *Aakasaka* Palace (a.k.a. Tokyo "State Guest House") where foreign kings and queens stay during state visits. He also learned how to safely build helicopter landing pads on the side of historically protected palaces. These types of projects are not undertaken by the U.S. government (Waller 2011, personal communication). Thus, Waller experienced an extraordinary level of access and was lent a unique perspective even from the aspect of the Mansfield Fellowship Program, which universally provides remarkable access due to the negotiation process undertaken by the Mansfield personnel and the National Personnel Authority (see chapter 1).

Third, given his own deep professional background, Waller developed a very clear set of goals that MLIT received with great seriousness and ensured that Waller's entire plan was completed in a clockwork-like fashion over his yearlong stay treating his MFP proposal as if it were a concrete itinerary. As Waller (2009, 14) explains,

> Even though I had only one placement, I had access to the numerous different programs that are administered by MLIT. In other words I had a short internship with the Urban Renaissance Agency (an urban planning program), attended seminars regarding Federal housing projects, and other area of interests not administered by my immediate division/office. . . . I was able to learn about Japan's innovations in seismic design and retrofitting of existing buildings, green building efforts (i.e., solar, wind, green roofs, and heat islands), historic preservation and the programming and planning for new buildings.

This is exceedingly rare for the Mansfield Fellows whose placements ordinarily require significant negotiations by both the NPA and the MFP with the hosting ministries and agencies to ascertain the appropriate policy network and necessary permissions from the host agencies. As noted previously, such as in the case of Jeffrey Seay and Yukiko Ellis who both adjusted their Fellowship goals upon arrival in Japan, this also requires flexibility on the part of the Fellows themselves due to unanticipated constraints or challenges which preclude completing their proposals precisely as planned.

Turning to Waller's primary and secondary goals is illustrative of the specificity with which he approached his Fellowship experience:

> My primary goals were to study Japan's approach to Universal and Barrier Free Design programs. Specifically I studied the following: (1) accessibility regulations [and] design guidelines. (2) Innovative design ideas for the disabled and elderly. My secondary goals were to study the following: (1) seismic design and retrofits, (2) historic preservation, (3) energy efficiency, [and] (4) lowering buildings['] CO2 emissions (Waller 2009, 1).

To accomplish these goals, Waller deeply assessed four government building projects. The paragraphs below briefly review two of his research locations at the Hamamatsu National Government Building and the Inoue Eye Clinic in Tokyo to illustrate the policymaking network of the Government Buildings Department, Architectural and Building Engineering Division of MLITT. Again, this Division would have previously fallen under the purview of the Ministry of Construction prior to its merger with the Ministry of Transport.

First, the Hamamatsu National Government Building was a new project intended to incorporate the latest in design concepts to service a variety of special needs in the general population that would be visiting the government's immigration office on the first floor, tax offices on the second and third floors, and probation operations on the remaining floors (Waller 2009, 2–5). In other words, the building was extremely multi-functional and thus had to be accessible for individuals with varying challenges and disabilities. As such, the "building's construction budget allowed for community involvement and many disabled organizations along with the designers and government representatives [to] create . . . a highly usable building for all" along with providing surveys to assess the building's amenities (Waller 2009, 2). To this end, a variety of innovations were made to the design to ensure as many items on the Universal/Barrier Free list were accomplished including, among other things:

1. "Alarm/panic buttons . . . in two locations, [since] many paralyzed wheelchair users fall when transferring from their wheelchair to the toilet, having an alarm/panic button near the floor is helpful to those who have fallen. This design practice does not exist in the United States."
2. "Open Legged Sofa/Seating [to] allow . . . the seeing-eye dog to sit under the sofa or seating. This is just common sense, however many individuals who specify furniture, rarely think of such things."
3. "Continuous Two-Level Handrails in Stairwells [to] provide assistance to those that are elderly or blind, as they serve as a guide. The two-levels assist those who are shorter and taller."

4. "Contrast between Walls and Urinals, [to] assist . . . those with low-vision, by assuring the user that they are standing in the right location."
5. "Oversized Elevator Call Buttons/Oversized Floor Buttons [to] help . . . those with arthritis [and/or] low-vision. Oversized Elevator Directional Notification, [to] assist those with low-vision and are helpful to the hearing impaired that cannot hear the elevator announcer. Side-Panel Elevator Floor Buttons [to] assist wheelchair users and children. This is not required in the United States."
6. "Large Contrasting Directional Signage, Pictograms and Information Boards, [to] assist . . . those with hearing disabilities."
7. "Restrooms [with] cane holders . . . in bright-yellow for the elderly and those with ambulatory issues. Not required in the United States."
8. "Flashing Red Alarms [to] notify the Hearing Impaired during times of emergency."
9. "[An] Area of Rescue was designed for wheelchair users or others that cannot use the stairs during times of emergency" (Waller 2009, 2–3).

Waller goes into significantly more detail in his final report describing the particular details and highlights of the building including the "Paraplegic Restroom" along with the "Ostomate Restroom," which incorporate intricate and deep thinking details to serve the needs of those with special challenges along with photos to demonstrate their designs. As Waller concludes, "I have never seen such a well-thought-out designed restroom in my twenty eight year career as a designer" (Waller 2009, 4).

The question naturally becomes how MLITT might achieve such intricate and thorough analyses and subsequently translate these findings into actual building designs? To best reveal how this model works, this analysis presents verbatim cites from a digitally recorded interview between Waller and the author (Waller 2010, personal communication) along with details from Waller's Final Report (Waller 2009).

> Here in the United States, we have the Americans With Disabilities Act and the Architectural Barriers Act. . . . The Americans With Disabilities Act is really a lot about civil rights legislation. . . . It's just not about design, it's about how we're going to treat people. . . . There is *some* form of design element . . . that goes with that. But in Japan, they take a totally different approach. . . . Every other year, they update their ideas about what would be the best design solutions for the disabled or the elderly. And so, there are not hard and steadfast rules like we have in the United States . . . (Waller 2010, personal communication).

Thus, design innovations are not made based upon legal strictures related to a legal rights driven analysis. Rather, the Government Buildings Department of the former Ministry of Construction focuses on the substance of the technical innovations. As Waller (2010, personal communication) explains, this process is more in depth than he perceives is possible in the United States based upon his extensive experience,

> [Waller] [W]e frequently think about commerce and business. And commerce and business want to know . . . "[I]s my bathroom door wide enough? Have I complied [with the law]? And if I have, just give me a check mark and I'm moving on." In Japan, there tends to be . . . [a] global understanding about . . . the trend of the population. *These* are the things we might need to do to change. Have we thought about a streetscape that doesn't have curbs because elderly people don't like that? [Instead,] have we thought about a new way we might have water drainage come down a new gully that is covered by a grid? I found that they approached problems totally different than we do here in the United States. . . .
> [The Government Buildings Department staff have the] continuous ability to study every detail and gather all the data points available. And nobody make[s] a decision . . . until all the data points [are] known and understood. But once we kn[o]w them and all understood them *and we all agreed*, then we [are] going to . . . go do *that*. . . . I don't know that we here in the United States could spend all that time and money to come to a decision . . . On a building they can gather CO_2 output, the temperature of day, the number of people, the lighting load—all of these endless details that we are here in the U.S. are starting to get our computers to do for us. But they . . . have all these people to gather and input this data and then figure out how to do it right. And then they will do prototypes, and then they will study that prototype . . . (emphasis added).

As this conversation reveals, to build this consensus requires heavy reliance on a data driven, multi-iterative problem solving process where policy choices are based on the greatest number of "data points" possible. At the Hamamatsu National Government building this also included "community involvement and many disabled organizations along with the designers and government representatives" not to mention a "day-to-day management team that has been trained to address the needs of the disabled" (Waller 2009, 2). Thus, numerous stakeholders intervene and the policy context is remarkable for the multiplicity of community principals along with local government agents involved in the policymaking process at various stages. This is the key distinguishing point from Ellis's experience at MIC in that participation is not confined to officials at the local government level in ongoing government-to-government (G2G) relations. Rather, the policy network in this case is open to various societal interests and players.

A second project which Waller observed up close also lends a clear picture of the array of research participants and stakeholders in government building project assessment. Specifically, Waller observed the process of subjecting the Inoue Eye Clinic project in Tokyo to a "total post-occupancy review" about two years after it had been completely renovated and modernized. To do so, the staff of the Clinic chose Sunday as the day for the review since the facility was closed on that day (Waller 2009, 6). Some of the participants included, "receptionists, nurses and three doctors . . . on site to participate in the review," "the elderly, hearing and visually impaired, color-blind subjects, wheelchair users, mothers with small children, and foreigners" (Waller 2009, 6). Waller himself was asked to serve as a research subject by viewing the signage and other features for foreigners visiting the eye surgery office to assess their accessibility and clarity for a foreign audience (Waller 2009, 6).

The question is thus where the opportunities for bureaucratic drift may arise given the multi-iterative and complex nature of the policy setting? Indeed, does this case study affirm the prediction that the presence of multiple principals increases the opportunities for bureaucratic drift with the causal effect of enhanced bureaucratic discretion? Once again, the answer is "yes." The voluminous data generated through the foregoing process requires significant teams of analysts at MLITT to sift through it in order to come up with concrete policy recommendations. As Waller explains, there was a clear division of labor between two separate groups in his division of MLITT what he calls the "Worker Bee Team" and the "Panel of Experts." As the interview continued,

> **Author:** I am really curious about this. If you are going to get those kind of holistic approaches to problem solving, how do you pull a team together that has the sufficient expertise to make it happen?
>
> . . .
>
> **Waller:** There were two [teams] that I dealt with. There was a team that was the "Worker Bee Team," if you would. And then, for example, in looking at innovative ideas, we would bring in what I would call a "Panel of Outside Experts," which we might call [here in the U.S.] a presidential commission . . . [T]hey would have . . . a chairman . . . [who] would be compensated . . . and some of [the expert panelists] were like really true in depth, good thinkers . . . And they would assess . . . what they thought. So, for example, the [Expert Panel] Team that I was working with had done several accessibility projects for the disabled and elderly and then they had designed the next questionnaire that they were going to be using this next year and they were looking at the pilots that they had accomplished and then had the research data to back it up . . . And this panel would meet for half a day, somebody would present the [data], and then they would go over it and then look at it and then discuss it and then come back maybe a couple of months later after the group had made some recommendations and they would try to implement it. . . . [A] lot of professors were . . . participating in what I would call the Japanese form of democracy . . . . We have subject matter expert panels here on certain things at GSA, but not on the level that they had.

> **Author:** Interesting. So, would they call in professors from various universities on an *ad hoc* basis as needed or was this . . . a very formalistic approach to interfacing with . . .?
> **Waller:** It's very formalistic . . . I found that professors were . . . your larger group of panel experts.
> . . .
> **Waller:** When it came to things such as earthquakes, most of the people that were presented to me and hung around me were people who had PhD's in seismology . . . or engineering. These were . . . individuals . . . that I would not normally come across in my daily job here at GSA (Waller 2009, personal communication).

As this excerpt of a much longer interview demonstrates, there are significant distinctions with the first case study of the SBJ in that the Government Buildings Department of MLITT invests significant time in constructing both a core of technically competent staff, an effective epistemic community for design projects, and has the discretion to determine how it will employ its significant community linkages with both local administrators and interested stakeholders. Individuals with significant technical experience and knowledge are called upon for insight at regular stages while their recommendations are experimentally implemented through a multi-iterative process. This permits numerous revisions to models and designs until as many facets of the problem have been addressed and errors removed. Only thereafter does a consensus ultimately emerge and the project move forward with actual implementation in the design of government buildings. Lest one might suppose that this process was only true for this particular project, Waller observed two other projects in two separate geographical locations with very similar approaches to seeking perfection in a design context.

Thus, the predictions of this project are affirmed with regard to the Government Buildings Department of MLITT. Once again, these theoretical observations are posited by the author alone and ought not to be ascribed to Waller. First, the presence of multiple principals in this case provides opportunities for bureaucratic drift. In this instance, MLITT has selected a multi-iterative, tradition bound decision-making process with various stages and teams. Thus, just as in the case of James Miller's observations of MIC in the context of radio wave regulation, the result is a hands-on approach to deciding the ultimate outcome and the process to reach that outcome. As such, the effect is to enhance the discretion of MLITT in this case to achieve outcomes in the manner it chooses.

NAVEEN C. RAO, ESQ.

As indicated in the opening remarks, the next case study at MLITT provides an interesting comparison to the Government Buildings Department. While also a policy context involving multiple principals, aviation with its heritage in the Ministry of Transportation diverges significantly from a department that traces its lineage to the Ministry of Construction. This clarifies why sectionalism has gripped MLITT as a whole giving it a schizophrenic appearance to the general public.

Naveen Rao also completed his Mansfield Fellowship (2003–2005) at the Ministry of Land, Infrastructure, and Transport (MLIT), but with the Japan Civil Aviation Bureau (JCAB) whose responsibilities had been under the purview of the former Ministry of Transportation (MOT) rather than the Ministry of Construction (MOC). By way of background, while at the time of this writing Rao is an associate at Jones Day, he served as an attorney at the U.S. Federal Aviation Administration (FAA) from 2001–2006 before serving from October 2006–April 2008 as the Counselor to the Assistant Secretary at the U.S. Department of Transportation. He earned his JD from the University of Arizona, James E. Rogers College of Law and MBA from the University of Arizona, Eller College of Management. He also had significant Japanese language experience prior to his Mansfield Fellowship serving as an Assistant Language Teacher (ALT) from 1995–1997 in Japan through the Japan Exchange and Teaching Programme (JET) under the auspices of Japan's Ministry of Foreign Affairs (MOFA). Rao completed a multi-placement Fellowship experience at the (then) Ministry of Land, Infrastructure, and Transport (MLIT) in the JCAB, before proceeding to the Japan Fair Trade Commission (JFTC) (Economic Affairs Bureau) and All Nippon Airways (ANA) (Rao 2005, 3).

Japan Civil Aviation Bureau and the Aviation Industries Division

The discussion below focuses on his six months of experience at JCAB first with the Flight Standards Section (September 2004–November 2004) and then the Aviation Industries Division (December 2004–March 2005). Rao's specific placement goals were to (1) "[g]ain an understanding of aviation regulation in Japan from both safety and economic perspectives" and (2) "learn how Japan applies antitrust policies in merger reviews." To this end, he spent two months in JCAB examining Japan's safety oversight regulatory system and then four months evaluating the office "responsible for developing and administering economic policy for the Japanese airline industry" (Rao 2005, 4).

Turning to his first stint in JCAB's Flight Standards Section understanding safety oversight, Rao interestingly discovered the *absence* of multiple principals. Given the technical nature of the responsibilities for airline safety, the Flight Standards Section was focused on the technical aspects of airline safety. As predicted, this meant that the Flight Standards Section had decreased opportunities for bureaucratic drift with the effect of decreased discretionary authority. This was the case as strict adherence to technical standards was expected. Even so, this discussion is important to clarify the distinct policy approach of the Flight Standards Section as it departs notably from the Government Buildings Department. This helps to clarify the different traditions of the two divisions. As Rao (2005, 3) explains in his Final Report,

> the US government (or perhaps my agency is particular) is organized around people or their skill sets. For instance, the Federal Aviation Administration (FAA) is divided into groups of people [with] similar backgrounds [who] are grouped together physically: the mechanics are in one area, pilots in another area, and attorneys in yet another. Whenever a project requires input and expertise from [a] specific area, teams are assembled by pulling people from the various groups and putting them together. In contrast, [JCAB] sets up task-centered offices dedicated to a specific purpose and then staffs them with the human resources required for the task.

Given this intriguing notation in his Final Report, the author asked Rao to clarify his meaning in an interview.

> **Author:** [Your Final Report mentioned that] Japanese agencies are organized around a specific mission whereas the United States government is . . . about the people's skill sets. So what happens in Japan when, say, a team finds out it needs help? So, in other words, what happens when a problem . . . requires outside assistance? How does the Japanese agency handle that kind of inter-agency cooperation? Do they have procedures in place . . .? What do they do when that team that is organized around that mission just can't . . . handle it by itself?
> **Rao:** I don't know about inter-agency so much as I know about intra-agency [cooperation]. . . . I don't know if you have heard the word, *takobaya*, it means "Octopus Room," . . . [or] what we call a "fishbowl" office. Basically, if there is a certain issue, they will pull people from different offices, from wherever, and throw them at a problem. The contrast I was trying to draw out in my [Final] Report, [is that] in the U.S., if something needs to be done, . . . it will go through multiple levels of review. . . . [In Japan, the Flight Standards Section's] approach is . . . issue-centric (Rao 2009, personal communication).

In other words, the policymaking approach of the Flight Standards Section was not of the same extensive and relatively congealed character as observed in MLITT's Government Buildings Department noted above. Rather, the context was far more ad hoc with the Section forming and then reforming teams in an issue centric fashion. As Rao continues,

> What they will do in concrete terms, for instance, is if they realize they have a problem (or want to try to solve a problem or anticipate one), . . . is look at what the U.S. has done, then figure out what the EU has done, and then figure

> out what they want to do using those two as a reference. That's to me a very rational, very deliberative process. So it is very non-political. The political considerations do not really interfere. There is no bargaining (Rao 2009, personal communication).

Further dialogue revealed that the Division relied upon the "Octopus Room" tradition of that particular agency. Specifically, this meant that the agency would regularly reach its tentacles out to pull staff in as needed to solve problems. Rao was surprised with the resulting dearth of subject matter experts vis-à-vis the FAA in the United States who ended up in the Octopus Room (Rao 2009, personal communication). When combined with the hidebound personnel rotation system, Rao (2005, 3) perceived that this led to significant challenges,

> The Japanese human resource system of rotating generalists throughout an agency is not suited for an area as complex and knowledge intensive as aviation safety oversight. In the JCAB Flight Standards section, most inspectors charged with airline operational oversight have engineering/technical school backgrounds and do not have pilot or mechanic qualifications or experience. The pilots in the section are limited to examining pilot proficiency and not evaluating broader safety matters. I believe this creates a serious problem. Further they are only kept in positions for relatively brief periods of time.

Again, this was unlike Waller's experience in the Government Buildings Department of MLITT, which was very well populated with highly qualified staffers or boards of experts from the epistemic community. Instead, JCAB appeared to struggle with the need for more expert staffers.

This begs the question of why the difference? Part of the answer appears to be based in sheer necessity as the administrators in the Flight Standards Section are dealing with the life and death issues of airline safety and thus must rationally assess problems while relying upon administrative procedures and protocols. Anything less than quick and straightforward evaluation of choices based upon those considerations might result in immediate safety issues which cannot be tolerated (Rao 2005). Thus, the immediacy of safety in the airline industry, unlike in the more experimental and longitudinal focus of government building design, may partially explain the difference in problem solving. However, part of the difference may also be due to the disparate policymaking approach of the aviation division vis-à-vis the buildings division due to their distinct institutional legacies and memories.

Specifically, turning to Rao's discussion of his next placement at MLITT reveals that when it comes to the actual creation of aviation policy, JCAB is involved in a highly contentious arena populated by multiple principals. This is dramatically different from the consensus driven policy context of the Government Buildings Department. Yet, despite this difference, this inquiry predicts heightened opportunities for bureaucratic drift with the effect of increased discretion for JCAB regulators due to the presence of multiple principals.

To begin, this policy arena is home to significant controversy over infrastructure projects. A lengthy citation from Rao's Final Report (2005) is particularly illustrative,

> There are new airports scheduled to open at Kitakyushu, Kobe, Shizuoka, and a new runway for Kansai International Airport is under construction. Under standard cost/benefit analyses, these projects are not economically viable or are marginal at best. Kansai airport actually is underutilized as it exists today with a single runway and a new runway will only saddle the operating company and airlines that use the airport with new costs. Nonetheless the Kansai airport runway and other projects are in various states of going forward. Since part of JCAB's mission is to ensure adequate air service around the country, they must 'encourage' the airlines to operate to these airports. . . .
> There is grassroots political pressure on both the airlines and the government to see these airports are served as soon as one is built. When an airline decides to suspend a route or reduce service, there is usually an outcry from the localities affected. I witnessed one poignant example while at JCAB in which a representative from a rural prefecture hauled out the MLIT Minister and the JCAB Director General to the Diet and grilled them on live TV simply because JAL [Japan Airlines] reduced their schedule from five flights per day to four. . . . The issue of air service really touches a nerve in rural areas. The Aviation Industries Division is charged with managing and balancing with all of these competing and conflicting equities. I think a casual observer may look at JCAB's activities in the domestic airline industry and simply conclude it is yet another example of crude government interference in the marketplace . . . . But I think that would miss the point.

This is the key point that provides support for the prediction that a policy context involving multiple principals—in this case politicians seeking to please various rural areas that wish to have air service—increases opportunities for bureaucratic drift. As Rao's description notes, the Aviation Industries Division was ultimately left with the managerial role of balancing the various interests at the table. Rao summarizes how this outcome may be understood in the broader context of democracy and the potential for bureaucratic drift,

> democratic, free-market societies gravitate towards equilibriums as defined by economic as well as political constraints. While the equilibriums may be suboptimal in an academic sense, I think they are optimal in that they reflect the actual tradeoffs that a society is willing to make. I think this is true in any democratic society. Once one recognizes that the priorities that go into the Japanese decision making process are different from those elsewhere and the scarcity of certain resources, things make some sense (Rao 2005, 5–6).

In other words, the existence of multiple principals enhances opportunities for bureaucratic drift resulting in the exercise of increased discretion as agents are left to make the tough choices where scarce resources are involved. This may indeed "irritate" the principals involved as none of them are entirely satisfied with the policy choices the agents have made (Ehrhardt 2009).

CASE STUDY 3: SUMMARY

As the foregoing case studies of Waller and Rao's experiences at MLITT confirm, the ministry today is internally divided. On the one hand, policy divisions that would have formerly fallen under the Ministry of Construction (MOC) operate in relatively technical contexts which involve multiple principals. This has meant reliance on a tradition of consensus building amongst the multiplicity of stakeholders through an extraordinarily formalistic and incremental process. Just as in the case of Miller's observations of MIC's Radio Policy Division, MLITT's Government Buildings Division exercises its discretion to adopt a deeply hands-on approach. Indeed, the foregoing reveals a carefully methodical, meticulous, multi-iterative policymaking process as various community stakeholders including the epistemic community are involved in building design, modernization, and renovation.

This example is even more illuminating when juxtaposed to the divisions that would have formerly fallen under the Ministry of Transportation (MOT). In the first instance, when it came to applying standards of aviation safety, multiple principals were not involved in policy implementation. This meant the lack of observed bureaucratic drift. Thus, the effect was the significant decline of administrative discretion as agents relied upon the Octopus Room model to cobble together ad hoc coalitions of agents to resolve problems. This was apparently due to the safety issues involved at this stage and the need for strict reliance to safety protocols and also apparently is a style of decision-making deeply rooted in this division.

However, turning to the policymaking level, multiple principals are involved in seeking to shape aviation policy. As predicted, this results in significant opportunities for bureaucratic drift with the effect of increased discretion for the implicated agents to manage and balance the competing interests. Indeed, the Aviation Industries Division is likely to continually irritate the various principals involved as it has the discretion to choose practical rather than political outcomes when selecting which runways to build or airline routes to fly.

In the end, MLITT's schizophrenic presentation to the public certainly creates the perception of bureaucratic drift. These cases reveal that administrative agents in both divisions are allowing their own preferences to determine the pace, style, and outcome of policymaking in both the buildings and aviation divisions. Again, the directive of political principals that sectionalism should be replaced by cooperation was very clearly articulated in public surveys as was the intent that these ministries should work together under a single roof. This was precisely the unifying logic of initial reform merging the two ministries and subsequently expanding the ministry's responsibilities to include tourism. The implicated agencies have clearly demonstrated, however, that where multiple principals are involved in the policymaking arena, opportunities for bureaucratic drift appear with enhanced bureaucratic discretion ultimately winning the day. The next chapter turns to the fourth case study presented by this project.

NOTE

1. Waller grew up in Montana having graduated with his BA in art (professional design with an emphasis in interiors) and thus was very familiar with the outstanding diplomatic career and service of Ambassador Mike Mansfield for whom the Mike Mansfield Program is named and thus had known of the MFP long prior to his participation. Waller also holds his master's in public administration from American University and a certificate in quality management. He has continued to retain his connection to his home state. He is one of the 2002 recipients of the Alumni Achievement Award from Montana State University, and in 2003 the Waller-Yoblonsky Gallery within the Melvin Graduate Studio Building/School of Art was named in his honor. Additionally, he served on the board of directors of Montana State University's Foundation (Waller 2011).

Chapter Five

Case Study 4

Ministry of Health, Labor, and Welfare (MHLW)

This case study of the Ministry of Health, Labor, and Welfare (MHLW) presents the Mansfield Fellowship experiences of Jim Kariya and Dr. Ken Kobayashi.[1] MHLW presents policy settings involving multiple principals which have created enhanced opportunities for bureaucratic drift. As expected, the effect has been the exercise of enhanced administrative discretion. The following first begins with an introduction to MHLW before turning to the Mansfield Fellowships of Kobayashi and Kariya. Finally, the lessons learned from this case study are summarized.

INTRODUCTION

The MHLW was created in 2001, two years after the National Government Organization Act, by merging the Ministry of Health and Welfare (MHW) with the Ministry of Labor (MoL).[2] The merger was the outgrowth of the final report of the 1996 Administrative Reform Council headed by the prime minister's office. As part of its expansive mission to restructure government for Japan's new future of an aging population, the Council concluded that consolidating certain key functions of government was essential to maintain the same level of service. Specifically, the Council determined that the national interest fell into four critical areas, (1) "[c]ontinuance of the [n]ation," (2) "[s]ecurity and expanding . . . national wealth," (3) "[s]ecurity and improvement of living conditions," and (4) "[s]uccession and maturing of education and national culture."

Interestingly, in some ways this move was dialing back the clock. The Ministry of Labor had been established in September of 1947 by endowing it with the responsibilities previously handled by the Ministry of Health and Welfare. However, today the recombined MHLW (2010a, 3) has an expansive mission,

> [T]o promote and disseminate science and technology in health, labor, and welfare administration in order to ensure good health and high quality of life with safety and comfort by utilizing the findings of state-of-the art research, and to safeguard people's daily lives by promptly taking measures against the threats to people's health. In order to accomplish these two big missions, the Ministry of Health, Labor, and Welfare holds jurisdiction over national research institutes and conducts [the] overall planning and coordination of science and technology.

Operationally, the MHLW had eleven bureaus and seven departments as of 2011, each overseeing significant and distinct policy areas. Examples of bureaus that would have been under the former Ministry of Labor include the Labor Standards Bureau, Pension Bureau, Employment Security Bureau, Social Welfare and War Victim's Relief Bureau, Equal Employment, Children, and Families Bureau. In contrast, examples of bureaus that would have fallen under the jurisdiction of the Ministry of Health and Welfare include the Health Insurance Bureau, Health Service Bureau, and Health Policy Bureau (MHLW 2010a, 2). Some departments also have responsibilities that overlapped between the two former ministries such as the Statistics and Information Department which houses the Policy Planning Division, Vital and Health Statistics Division, Social Statistics Division, Employment Statistics and Information Department Division, Wages and Labor Welfare Statistics Division (MHLW 2010a, 2). MHLW also has numerous affiliated institutions such as national research institutes along with various councils, regional bureaus, and external bureaus (MHLW 2010a, 2). This review illustrates the far reaching jurisdiction of today's MHLW and its direct impact on the lives of the Japanese people. For instance, in the aftermath of the March 2011 earthquakes and tsunami, the MHLW was deeply involved in managing many aspects of the government's response including medical relief efforts and inspecting food products to ensure their fitness for consumption. Thus, any deficiencies in the MHLW's performance are intimately felt by the Japanese people.

Various administrative problems were in fact manifested at the start of the decade of reform to centralize, streamline, and improve the governmental services managed by the MHLW. As with the internal divides within the MLITT considered in chapter 4, there are similar cleavages within the newly constituted MHLW,

> A similar in-house schism was observed in the Health, Labor and Welfare Ministry at the time of the medical system reform [in 2001]. Officials from the former Health and Welfare Ministry reportedly complained about the tardiness of those from the former Labor Ministry in the new ministry's attempts to avert the government's reform efforts, for example[,] by trying to win over Diet members. The organizational clout of the giant central government offices created through the mergers apparently has failed to match their size. [Public Management Minister Toranosuke] Katayama aptly compared the giant ministries to big ships: "Gigantic ships can only change their course slowly" (*The Daily Yomiuri* [Tokyo] 2002).

The resulting failures of the giant MHLW over the next decade reflected far more than administrative difficulties such as problems in effectively managing the pension records system, to quickly diagnosing patients and advising the public during outbreaks of the avian flu, novel influenza, and drug induced hepatitis (MHLW 2010b). Perhaps even more disturbing was the lag in the approval time of innovative drugs to combat new strains of infectious disease coupled with an ineffective national distribution system even for approved medications (MHLW 2010b). Clearly, bureaucratic drift had reached a zenith.

Simultaneously, the ongoing public clamor for change engendered a cyclical raft of attempted reforms. These calls led to various changes such as measures to provide pharmaceuticals more promptly via the "5-Year Strategy for Creation of Innovative Pharmaceuticals and Medical Devices" (MHLW 2007, 3). Some of the measures included (1) doubling the number of pharmaceutical reviewers to improve quality by adding 236 staff members over 3 years, (2) clarifying the review processes and standards, (3) establishing guidelines for global clinical trials, (4) considering the introduction of parallel scientific advice programs for global clinical trials among Japanese, American, and European regulators, and (5) promoting the rationalization and simplification of clinical trials along with the approval review process to ensure the safety of new medical devices (MHLW 2007, 3).

Even so, reforms came more slowly than expected. In a statement released by the MHLW in its 2009–2010 Annual Report, the ministry recognized the continued urgency of reforming its "Bureaucratic Culture." This is entirely in tune with the prediction of this project that bureaucratic drift is enhanced in policy contexts involving multiple principals. In this case, bureaucratic drift had a decidedly negative impact on the quality of its services. The MHLW (2010b) even went so far as to apologize formally to the Japanese public for its deficient performance,

> The Ministry of Health, Labour and Welfare is the mainstay of the Administration that protects the people's lives, and the people highly rely on it. However, the people have lost confidence in the Ministry . . . We deeply regret and

> sincerely apologize for destroying the foundation of the reliance because of the problems that have disappointed the people. . . .[We will seek reforms] to cut . . . red tape, stop wasting [resources,] and [improve the] pension, social insurance, pharmaceuticals[,] and health[care] [systems].

To open up the black box of MHLW and thereby understand these administrative challenges, the following considers the experiences of two Mansfield Fellows who completed their stints at MHW and MHLW prior to and during this merger, including Jim Kariya and Dr. Ken Kobayashi, respectively. These studies are supplemented with outside evidence on the regulatory arena of the healthcare industry in Japan. Doing so reveals a policymaking milieu similar in many ways to processes observed in the Ministry of Land, Infrastructure, Transport, and Tourism (MLITT) in chapter 4.

JIM KARIYA

Jim Kariya completed his Mansfield Fellowship (1996–1997) in the Ministry of Health and Welfare (MHW), Office of Environmental Chemicals Safety, and Food Chemistry Division. As an environmental scientist with the U.S. Environmental Protection Agency (EPA) in the Office of Prevention, Pesticide Programs and Toxic Substances and Office of Science Coordination and Policy, he was interested in evaluating Japan's risk assessment procedures and methodology (Kariya 1998, 1). He sought to determine,

> If there are more efficient ways to assess risk than what we use in the [United States], or if there are enough similarities between the two countries that risk assessments can be shared rather than duplicated in each place. . . [U]nderstanding how decisions are made in Japan about chemical risks might help us understand Japan's concerns in international discussions about priority-setting and control of joint problems (Kariya 1998, 1).

Notably, while Kariya was involved in a variety of "extra-ministerial activities" visiting the Diet, the judiciary, the Ministry of Finance (MOF), and attending various other related functions, he completed his year-long placement at a single ministry similar to Waller's fellowship experience (see chapter 4). This was justified because while there were various offices associated with non-pesticide chemical risk assessment, for his purposes "a large part of the human health risk assessment is done in . . . the Ministry of Health and Welfare" (Kariya 1998, 1–2).

Division of Chem-Bio and Division of Risk Assessment

Kariya found a key distinction between the research and policymaking offices of MHW. With regard to research affairs, MHW oversaw semi-autonomous national institutes. These included the National Institute of Health Sciences (NIHS) where Kariya spent several months first with the Division of Chem-Bio Informatics which was "involved with the collection, analysis, and dissemination of information related to chemical safety" along with other responsibilities related to data collection such as maintaining the Institute's library (Kariya 1998, 2). The division has since been renamed the Division of Biological Chemistry and Biologicals and continues to be responsible for a wide array of biopharmaceuticals ranging from growth hormones and monoclonal antibodies to biological products (MHW 2011). The division's duties are thus highly technical as it must test these various products for safety and efficacy. To do so, this involves "experimental preparations, the development of methods for their characterization (in terms of their structure, physiochemical properties, immunochemical properties and biological activity . . .) . . . as well as in viral safety evaluation, the standardization of analytical procedures for quality control," and establishing standards of test methods to be employed in the field (MHW 2011).[3] In short, feasibility and safety studies are heavily driven by data analysis and the procedural requisites of the field.

Kariya then moved into the Division of Risk Assessment, which studies "toxicological information related to 'new chemicals' (chemicals not yet in the marketplace), participates in international efforts to characterize the harmfulness of high-production-volume chemicals, and has general responsibilities for chemical risk assessment" (Kariya 1998, 2). In this capacity, at his first office placement, he became involved with developing the global website entitled Global Information Network on Chemicals (GINC) for the purveyance of chemical safety information and to help establish an international epistemic community in this arena (Kariya 1998, 2–3).[4] In his second assignment at the Division of Risk Assessment, he similarly became steeped in the intricacies of Japanese laws governing chemical risk assessment, while evaluating the procedures developed to implement these laws (Kariya 1998, 3).

Up to this point, therefore, Kariya's experiences were quite similar to those of Rao in aviation safety at the Japan Civil Aviation Bureau (JCAB) in that multiple principals were largely absent. Consequently, as predicted, the Division of Chem-Bio and Division of Risk Assessment manifested decreased bureaucratic drift with the effect of declining discretionary authority as they were expected to apply the requisites of Japanese laws along with evaluation procedures.

Food Chemistry Division

However, just like Rao's transition from aviation safety to the aviation industry division, upon moving from the Division of Risk Assessment to the Food

Chemistry Division, which is where he completed the majority of his Fellowship, Kariya stepped into a lively regulatory arena in which a variety of policy actors including the Ministry of Finance along with multiple principals including the judiciary and Diet played important roles. As he clarifies, "[l]earning about the context in which the Ministry of Health and Welfare worked was as important to understanding decision-making within the Ministry as learning about the Ministry itself" (Kariya 1998, 5). In fact, he found that several ministries influenced the regulatory process although "there seem to be fewer people involved with coordination of efforts across organizational lines, an activity in which many people are involved in the U.S." (Kariya 1998, 8).

Turning first to the Ministry of Finance, Kariya was impressed by its authority "to serve the same role as the U.S. President's Office of Management and Budget [OMB] preparations, with extensive powers to approve specific plans of every other Ministry . . . [although] oversight extends only so far: once monies are approved, there seem to be no accounting later" (Kariya 1998, 6–7). Thus, the MHW shared some power with a "sister" agency. Considering the judiciary second, Kariya observed a surprising lack of formulaic oversight by attorneys. In contrast, in the United States,

> Attorneys are an integral part of decision-making and policymaking . . . In Japan, there appears to be much less need for a review of actions by attorneys. There seems to be less discussion than in the U.S. about what actions a particular law requires and less concern about interpretation of ambiguities. . . . The cumbersomeness of the legal system in the U.S. was conspicuous by its relative absence in the Japanese system (Kariya 1998, 6).

In other words, attorneys seemingly did not add "many layers of oversight and quibbling over words" (Kariya 1998, 6).

With regard to the third key actor, the Japanese legislature (Diet), while Kariya perceived that it was not as powerful as the U.S. Congress in terms of regulatory oversight, nonetheless representatives of the ministries were continuously called upon to testify before the Diet. In fact, the Diet regularly made late night requests for information from ministry officials (Kariya 1998, 6). As Kariya (1998, 6) explains,

> Given the weak position of the Diet, it was somewhat surprising to see the importance that the Ministries place on responding to inquiries . . . This system seems to cause legions of government bureaucrats to stay late every night merely on standby to answer questions that might arise from Diet members. The system is all the more surprising in that the result is a scripted exchange between the Diet and the Ministry, with all questions known in advance and all answers carefully read in response. There seems to be little chance for cross-examination or follow-up, which is so important to fact-finding and opinion-development in the U.S.

Naturally, a scripted performance does not provide room for policy discussion and debate, which is the central characteristic of a U.S. congressional hearing. Fact finding, in other words, is infeasible if the work of government is a formulaic series of questions and answers without concomitant dialogue.[5] Thus, these briefings are more along the lines of pro forma demonstrations than an exertion of power by Diet members (see chapter 1 for a description of the powers of the prime minister). At the same time, perhaps as a result of this continuous, largely scripted dialogue, the regulatory network remained relatively closed to the general public,

> There is a formal method of publication of governmental decisions, somewhat similar to our Federal Register, but in contrast to the U.S. system there seems to be little or no publication of the records that led to those decisions. I saw few if any supporting documents of the sort that the U.S. EPA routinely prepares and releases for public comment on specific cases, and little deliberate involvement of "stakeholders" in developing policies or procedures (Kariya 1998, 7).

Kariya notes that this system was under review for possible changes to permit greater input from outside societal forces.[6] The consequence for the relationship between the ministries in this policy context and the general public is profound. As Kariya (1998, 6) summarizes, while he perceives that this may have the tendency to enhance the quality of decision-making from his insider's perspective,

> [w]hat might be seen as efficiency by some . . . may be seen by others as a potential for abuse of discretion. With the ability to draft laws and with little oversight either by the Diet or by the judiciary, it is unclear that the[se] Ministries need to be responsive to an electorate or any other outside influence that may disagree with them.

As this discussion clarifies, both predictions of this project are confirmed in this case. Indeed, various interests were irritated by the continuing problems attached to bureaucratic drift which continued to attenuate the linkage between bureaucracy and democracy. Thus, Kariya sets the stage for the next participant-observer below who completed his Fellowship as the MHW became the MHLW.

KEN KOBAYASHI, MD

Dr. Ken Kobayashi completed his Mansfield Fellowship (1999–2001) at the Ministry of Health and Welfare (MHW) at the time of its merger with the Ministry of Labor (MoL) in the Evaluation and Licensing Division of the Pharmaceutical Safety Bureau before proceeding to complete a placement at the National Cancer Center. At the time of his Fellowship, Kobayashi was a Medical Officer (GS-15) with the U.S. Food and Drug Administration (FDA) in the Division of Oncology Drug Products where he reviewed and monitored clinical trials of anticancer drugs. Additionally, Kobayashi was an attending physician at the National Naval Medical Center, and a Clinical Assistant Professor of Medicine at the Uniformed Services University of Health Sciences. He served on numerous committees and coauthored twenty-six articles and book chapters at the time of his fellowship application, along with serving as the editor of numerous professional journals. Finally, Kobayashi is board certified in internal medicine and medical oncology, and board eligible in clinical pharmacology with licenses to practice medicine in Maryland and Illinois having earned his MD from Northwestern University in 1986.

Evaluation and Licensing Division of the Pharmaceutical Safety Bureau

The importance of this rendition becomes apparent when we learn that just like Ellis' favorable reception at MIC as a representative of the U.S. Census Bureau, Kobayashi was welcomed into the MHW.[7] First, the U.S. Food and Drug Administration (FDA) wielded considerable clout in Japan. Second, Kobayashi is clearly a senior expert in his field who came with sufficient experience to suggest he would be a helpful ally vis-à-vis future bilateral relations. Third, by his own assessment, Kobayashi also had the added benefit of looking and sounding Japanese due to his background as ethnically Japanese and thus adjusted readily to the workplace and policy context. Fourth, family and friends in Japan were well-situated to assist in the placement process with the MHW. Fifth, Kobayashi's timing was also opportune as this was a significant period of regulatory change not only of the healthcare sector in general, but also specifically in the MHW, which merged with the MoL to become the MHLW by the next year. Therefore, Kobayashi's MHW hosts perceived that Kobayashi might serve as an agent for positive regulatory change. As Kobayashi explains, "They wanted me to take back as much information as I could . . . to the FDA. They wanted the FDA to understand them . . . and understand the weaknesses of the Japan system so that the FDA would pressure the Japanese government to change" (Kobayashi 2010, personal communication).[8]

The driving impetus behind the intended changes was the concern that the regulatory structure in place did not permit the quick evaluation, approval, and importation of better drugs into Japan to best serve the national interest.

In fact, the year after Kobayashi's Fellowship in July 2002, the Diet enacted the Pharmaceutical Affairs Law (PAL) which would be fully implemented by 2005. Among other things, PAL sought "to fortify . . . the review system for approval and license [of new drugs] while taking international conformity into account" (Gross and Tran 2003). This was accompanied by significant reorganization of MHLW's policy network including "establishing an independent administrative institution . . . under the auspices of the National Institute of Health Sciences—and the Japan Association for the Advancement of Medical Equipment (JAMME). . . . to ensure the quality, safety, and efficacy of medicines and medical equipment in Japan" (Gross and Tran 2003). Even so, by 2008, the need for a regulatory overhaul remained. As one source explained, although the Japanese pharmaceutical market ranked second in the world at nearly 8 trillion yen, Japan's drug approval process remained in need of improvement,

> While clinical studies can vary greatly in length, the average duration of a clinical trial in Japan is approximately four years. This compares with just 18 months in the U.S. and the U.K. and 30 months in France—a phenomenon known locally as "drug lag." The government has been facing growing criticism for the length of the clinical trial stage in Japan, and the MHLW has recently released several new policies to address these concerns and accelerate the process, including the hiring of additional staff (Nagasaka et al. 2008).

The ineffectiveness of the FY2008 reforms were confirmed by the MHLW in its 2010 Annual Report (MHLW 2010b). As part of its overall review of its "framework of policy evaluation," the MHLW specifically found that it faced a budget shortfall that was interfering with its ability "[t]o offer new pharmaceuticals [and deliver] medical devices promptly with high efficiency" leading to continual "drug lag" and "device lag" (MHLW 2010b). Thus, the MHLW requested budget adjustments to permit more effective management of these decisions. In personal communication, Kobayashi notes that many of these and other significant efforts to modernize the regulatory framework are the consequence of pressure from the corporate world, media, and senior level administrative officials who recognized the need to reform rather than from outside pressure exerted by the U.S. FDA (Kobayashi 2010, personal communication).

Returning to Kobayashi's fellowship, as the foregoing review suggests, reputation (i.e., both that of the FDA and Kobayashi), familiarity with Japanese culture, "connections" (i.e., *nemawashi*), and the timing of his Fellowship all coalesced to ensure that Kobayashi achieved excellent access to his intended network at the start of a lengthy period of significant regulatory reform in drug regulation. Kobayashi confirms that this helped to "unlock all the doors" ensuring he "accomplished all [he] wanted and more" (Kobayashi 2010, personal communication). Again, accomplishing a Mansfield Proposal precisely as originally intended is unusual for the Fellows described above, with the exception of Anthony Waller (see chapter 4), as most Fellows typically had to be flexible based upon the actual policy context upon arriving in Japan.

In fact, a sixth factor that also enhanced access to this network might be termed "packaging." Kobayashi's Mansfield Fellowship Program placement proposal itself, developed in coordination with his supervisor at the FDA, sought to enhance bilateral understanding between the United States and Japan on "post-marketing drug safety monitoring system and implementation of the [International Conference on Harmonization or] 'ICH' guidelines in cancer clinical research" (Kobayashi 2001, 31). As his placement proposal clarifies,

"The central thesis is that the lessons learned will not only help strengthen the practice of clinical trials in drug development, but will also help FDA and Japan evaluate new anticancer drugs better. The objectives complement and deepen other FDA U.S.-Japan efforts to:

1. "Identify strengths and limitations of Japan's clinical trials system with respect to oncology drug development by studying Japanese trial design, execution, and data analysis."
2. "Study how drug regulators, clinical investigators, academic institutions, cooperative research groups, and pharmaceutical companies interact in designing, conducting, and interpreting cancer clinical research in Japan."
3. "Study Japan's post-marketing adverse events surveillance system and identify features and principles that might be incorporated into the U.S. system."
4. "Consolidate and broaden [a] network of contacts to include Japanese clinical investigators who can serve as resources for the Division of Oncology Drug Products and expand already established linkages with Japanese regulatory officials into the clinical arena."
5. "Establish mechanisms to facilitate dialog between Japan and [the] FDA on important issues of common interest" (Kobayashi 2001, 31).

As these five objectives reveal, Kobayashi is a highly competent source of information for this project.

Indeed, similar to Kariya's observations, Kobayashi's fellowship reviewed both the clinical research infrastructure and evaluative procedures related to clinical drug development and drug oncology as well as becoming acquainted with the subsequent policymaking and regulatory context for actually approving a new drug (Kobayashi 2010, personal communication). The discussion that follows considers Kobayashi's observations of the review and approval process in the context of a specific drug. As predicted, this review affirms significant bureaucratic drift with the effect of enhanced discretionary authority exercised by MHW agents.

From 1987 to 2002, the regulatory process for seeking the approval of new drugs in Japan was complex and underwent a variety of revisions due to complaints that the process was too slow. In short, the MHW had tasked various independent regulatory agencies with reviewing applications for new medical devices such as the Japan Association for the Advancement of Medical Equipment (JAAME) and the Pharmaceutical and Medical Device Examination Center (PMDEC). One significant source of drug lag came when the jurisdictions of these agencies overlapped leading to redundant layers of review (ITA 2000).

By the time Kobayashi had arrived in Japan in 2001 for his joint placement with PMDEC and MHW, this jurisdictional difficulty had been resolved. Specifically, the year before MHW had determined that PMDEC would be given full responsibility for about 10 percent of the yearly applicants including all those for "new devices" and "improved or modified devices" while leaving the remaining 90 percent of the applications for "me-too" devices (i.e., new brands of the same products) to JAAME (ITA 2000). Thus, the regulatory framework in which Kobayashi was involved, included a complex system of interactions between the PMDEC, expert panels of various industry, consumer, and academic representatives, and the MHW (see figure 5.1).

To break down the policymaking process, as figure 5.1 clarifies, initial applications for new drugs were first received by PMDEC, which had been tasked by the MHW with these duties. PMDEC would then undertake a series of drug evaluations in a multi-iterative process of consultations with various expert advisory panels and regulatory bodies such as the Organization for Pharmaceutical Safety and Research (OPSR). This level of review is represented in figure 5.1 by the arrows pointing to and from the box labeled "PMDEC" to the "Experts' discussion/Advisory experts" box." Hereinafter this level of review is called the First Tier. Frequently, a representative of MHW would be present at the First Tier of review, but only on as an observer. Once this stage of review was passed by an affirmative decision that the drug should be recommended for marketing in Japan, the drug moved into a second layer of review by the MHW itself. The MHW oversaw its own expert committee process represented in figure 5.1 by the box labeled "The Evaluation and Licensing Division, MHW" with arrows running to and from two Control Pharmaceutical Affairs Council (CPAC) committees including the regular and executive level committees. Hereinafter, this level of review is called the Second Tier. This process culminated in a final vote of whether to approve the drug or device and then the decision was made by MHW whether to license the drug or device. After thus passing both tiers of review, a drug would be approved or not for distribution.

With this system in mind, we may turn to a specific example of this review process that Kobayashi observed intimately both at the First Tier and

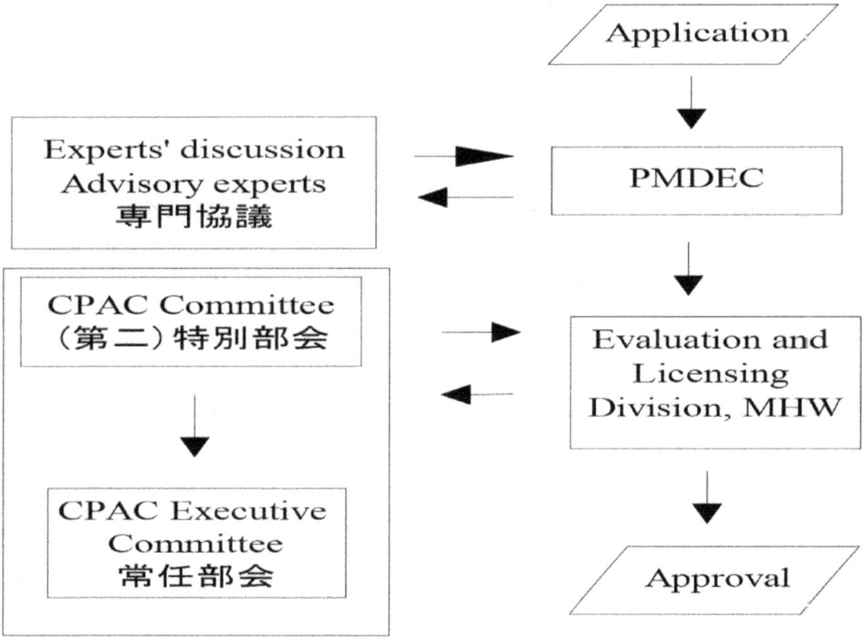

Figure 5.1. MHW-PMDEC 2001. Procures for Drug Approval and License (Kobayashi 2000b, 34). Created by the author with reference to: Kobayashi, Ken. "Figure 1: Procedures for Drug Approval and License" (Abridged from figure 4, Pharmaceutical Administration and Regulations in Japan [2000–2008], JPMA). National Personnel Authority of Japan.

Second Tier of review as he had the unusual opportunity to enjoy a joint placement with both PMDEC and MHW that facilitated access to each of these processes. To provide the appropriate context for the following discussion, Kobayashi describes the approval process of a cancer drug for treating hypercalcemia, which is the excess of calcium in the blood due often to malignancy or primary hyperparathyroidism (i.e., overactive parathyroid glands) (Hemphill 2011). In the particular case that Kobayashi observed, a U.S. corporation via a Japanese subsidiary sought licensure to sell a hypercalcemia treatment in Japan. Although this corporation had many licenses in other international markets, seeking licensure in Japan proved to be extremely challenging.

The substantive root of the problem was ascertaining the acceptable dosage of the medication in the Japanese marketplace. With regard to U.S. standards, the FDA had approved a higher dosage of the medication for use than was approved by the MHW in Japan thus raising significant regulatory and health concerns regarding the effect that the higher dosage would have when ingested by the public. The ordinary approach for overcoming these concerns is to conduct studies of the drug and, as Kobayashi notes, in Japan this is frequently done by "conducting a small (usually noncomparative) study using only the investigational agent and using the data derived from that so-called 'bridging study' to link the data from studies conducted outside Japan to the Japanese population" (Kobayashi 2000b, 13).

In this case, the U.S. firm interestingly sought a more informative and seemingly desirable approach of launching an intensive comparative study of the drug by "comparing the investigational agent with the drug currently approved in Japan for this indication, *apparently* with the intention of launching a development effort [of the drug] specifically for Japan" (Kobayashi 2000b, 13, emphasis added). The qualifying language of "apparently" is worth noting as the true intentions of foreign firms seeking to sell drugs in Japan may be unclear. As Kobayashi clarifies, "from the Japanese point of view [foreign firms may be] submitting applications and proposals as tests to see how the Japanese government reacts, with the intention of gathering data on which to base subsequent qualitative decisions regarding future drug development in Japan" (Kobayashi 2000b, 14).

The reception that the foreign firm received at both tiers of this review process was telling. Starting with the First Tier at the OPSR/PMDEC level of review, Kobayashi perceived from a professional standpoint as a physician and researcher that the reluctance of regulatory officials on the Japanese side was not grounded in evidentiary concerns. As Kobayashi explains, the foreign sponsor of the drug was relying upon a significant body of data, particularly on dosing decisions, due to the impressive size of the development effort of the drug abroad and thus "convincingly demonstrated that a lower dose than the proposed was likely to be inefficacious" (Kobayashi 2000b, 14). Likewise, language was not a problem as the foreign sponsor had retained a Japanese firm to assist throughout the regulatory process (Kobayashi 2000b, 12).

Instead, Kobayashi posits a variety of possible rationales other than a concern for substantive problem solving or scientific evaluation or even adherence to procedural rules as potentially driving the First Tier of review. First, he notes that many of the officials at this meeting lacked the requisite training as none of them were physicians, but rather pharmacists or pharmacokineticists. Although these individuals had conducted "truly impressive" efforts to gain the requisite knowledge through self-study, Kobayashi notes that

> drug development is a sufficiently difficult task that it is nearly impossible to obtain the necessary background and experience simply from reading. It is crucial to observe the outcomes of one's recommendations. Only in this manner can successful strategies or inappropriate advice be identified and, more

> importantly, the reasoning behind the errors be understood (Kobayashi 2001, 9).

However, even had the problem of unqualified staff been solved, Kobayashi also noted that the time allotted to evaluate the drug was insufficient. Kobayashi (2001, 9) explains this distinction in a lengthy comparison with the process in the U.S. Food and Drug Administration (FDA) at the time excerpted here,

> As an example, based on the minutes of [one] meeting, upcoming consultations for 28 companies were discussed in a 2-hour meeting [and at another] meeting, consultations for 19 companies were discussed in the same 2-hour period. To put these into perspective, the OPSR consultations appear intended to serve the same purpose as, for instance, an FDA-industry end of phase II meeting, in which advice is given to a company on, for instance, the appropriateness of a projected development plan, the appropriateness of various proposed clinical trials within the development plan, the appropriateness of preclinical development efforts, and so on. Typically, the company will submit a series of questions in advance of the meeting, which are then discussed and responses drafted individually between the primary reviewers within the concerned disciplines (for instance, medical, statistics, pharmacology/toxicology, clinical pharmacology, and chemistry/manufacturing) and their respective team leaders. These draft responses are then reviewed and revised as appropriate in a group meeting with all disciplines concerned, the Division Director of Oncology Drug Products, and other appropriate staff. The effort involved in researching the data, cases, and policies relevant to these questions and the subsequent drafting and meetings can easily consume several hours of each individual's time.

These problems were then further compounded as the involved officials simply lacked the resources to follow up and comment upon issues not directly raised by the affected sponsor company. Kobayashi also had the opportunity to observe the mode of interaction in periodic joint committee meetings held between OPSR and PMDEC with a representative of MHW present. Even those discussions seemed to remain focused on basic policy issues. As Kobayashi explains, "[i]n one instance in which a basic policy issue was challenged in favor of a more scientific approach, the OPSR senior staff unilaterally decided upon the advice to be given the company" at the First Tier of review (Kobayashi 2000a, 6). Clearly, this policymaking context exhibits significant discretionary authority vested in the agents.

Turning next to the Second Tier of decision-making, Kobayashi describes an even less scientific or data driven process. Specifically, unlike the expert committees and meetings run at the First Tier which permitted the involvement of various staff members, Second Tier meetings were closed, the members comprised of physicians and laymen, and the discussions run by the

Evaluation and Licensing Division of the MHLW (Kobayashi 2001, 10). Although PMDEC and OPSR were permitted to attend, OPSR staff did not do so on the occasions Kobayashi observed and the PMDEC staff did not take an active role in the discussions rather having previously prepared remarks read to the attendees without providing handouts or PowerPoints for reference (Kobayashi 2001, 10). Despite the presence of physicians, however, from a substantive, drug evaluative point-of-view, Kobayashi discovered that the discussion at this stage was even less scientifically driven than at the First Tier. Indeed, the PMDEC observers noted that although two of the four drugs under review by the committee on that occasion were anticancer agents, the committee did not include an oncologist. Rather, the issues addressed centered on more ancillary concerns that dealt with marketing and other concerns such as the labeling and color of the product. Kobayashi was particularly intrigued that the greatest amount of discussion time focused on whether it was appropriate to provide one of the agents under consideration in capsule form. As Kobayashi (2001, 11–12) explains,

> One of the agents under discussion is in a capsule form, and by far the greatest amount of discussion . . . centered on whether or not patients were likely to bite, rather than swallow, the capsule. To the best of my recollection, this consumed, in various guises, about one hour of a four-hour session. It should be stressed that the discussion did not advance to the level of, for instance, what the pharmacokinetic or clinical implications of chewing rather than swallowing a sustained-release capsule might be, but rather simply whether this was likely to occur. The question was raised by the consumer representative, who seemed to have little understanding of the scientific issues presented by the application.

Kobayashi's subsequent discussion is perhaps even more revealing as he initially considered that perhaps the lack of substantive discussion at this level of evaluation was simply because they had already been conducted elsewhere. As he explained,

> It might be speculated that the *nemawashi* phenomenon . . . in which all decisions are made in advance, outside the conference room, leaving room only for formalities at the time of the actual meeting, might account for the level of discussion in these meetings, but judging from the level of stress and intensity of preparation . . . this is doubtful. The level is similar to that on the part of FDA reviewers prior to an ODAC, and it is not credible that a meeting called simply for the sake of formality would engender that level of near-hysteria. Furthermore, having witnessed meetings that truly were the result of the *nemawashi* process . . ., it seems highly doubtful that these meetings were held purely for the sake of form. . . . the committee chair [received] an intensive one-hour briefing by the PMDEC review staff immediately prior to the start of the meeting on the major issues presented by the applications. While I

> was not present during this briefing, it seemed to have made some differences in the way he directed the meeting.

Thus, Kobayashi concluded over the course of his stint with both the MHW and the PMDEC that "[t]here is an interesting division of labor within the Japanese government. The *Shinsakanrika* (Pharmaceutical and Safety Bureau) at the MHW is mainly responsible for final decision-making and most discussions evaluate political, strategic, and policy-level issues. There is very little discussion of the data. These discussions seem to take place in the Organization for Pharmaceutical Safety Research (OPSR) and the Pharmaceutical and Medical Devices Evaluation Center" (Kobayashi 2001, 2).

As this rendition reveals, similar to the observations of Kariya above, the same dichotomous decision-making context was clearly visible in each of these two tiers of review. As Kariya explained in the context of chemical safety review, the First Tier of review was thus undertaken with an eye towards the data as the safety of the public depends upon the careful assessment of the new drug. As he explains, "In the review period at the [First Tier],[9] . . . there is a lot of back and forth, questions and inquiries posed to the sponsor [of a new drug] that each sponsor has to respond to . . . and there is a fair bit of bargaining and negotiation at that phase" (Kobayashi 2010, personal communication).

However, once the data has been gathered and the packets of information gathered, the next stage of review at the MHW clearly reveals both the presence of multiple principals and enhanced opportunities for bureaucratic drift with the effect of heightened administrative discretion. As Kobayashi (2010, personal communication) continues,

> But, when the application moves to the *Shinsanrika* [Pharmaceutical and Safety Bureau, MHW], then it is less about the data, and, I think, more about gaining consensus with the various stakeholders, whether they be internal or external. . . . [T]here is a lot of negotiation and bargaining and that kind of activity that goes on between the sponsor (corporation), between the *Shinsanrika*, between other stakeholders like . . . medical associations and so on, before an action is taken.

Importantly, Kobayashi (2010, personal communication) perceived that to some extent, this was inevitable in the medical regulation regardless of the national context,

> At the policymaking level, . . . [i]t's where science hits public policy. [T]here's always going to be a mixture of the two. . . . Depending on whether you are talking about a specific issue—action on the application or a broad policy decision—one or the other of those will dominate. . . . Whatever it is, it . . . will have elements of both because that's the job of a scientific agency.

Nonetheless, the extremes he observed in this tiered review process were remarkable both for their lack of substantive evaluation of the data by qualified experts and for the process itself. As he notes, in comparison to the processes at work in the FDA, "it is indeed a strange spectacle to witness this relatively amateurish process" (Kobayashi 2001, 12). Thus, the negative policy consequences for healthcare policy in Japan were a direct result confirmed by the public's irritation at bureaucratic drift and repeated attempts to reform the ministry.

CASE STUDY 4: SUMMARY

As this case study of the Ministry of Health, Labor, and Welfare demonstrates, this policy context implicates multiple principals involved in various stages of the policymaking process. Likewise, this case study affirms that bureaucratic drift with the effect of enhanced bureaucratic discretion is pervasive in this policy setting. Finally, perhaps unsurprisingly given Ehrhardt's (2009) study, despite numerous attempts at reform, this case evidences increasingly irritated principals with the performance of the Ministry in matters of significant social import from drug lag to approval of needed new pharmaceuticals. Chapter 6 below now turns to the final case study on the Ministry of Economy Trade and Industry (METI).

NOTES

1. For more information on the ministry, visit MHLW's website at http://www.mhlw.go.jp/english (accessed August 30, 2017).

2. Visit http://www1.mhlw.go.jp/english/wp_5/vol1/p2c7s4.html for more details (accessed August 30, 2017).

3. Visit http://www.nihs.go.jp/english/activities/dbcb.html for more details.

4. The GINC lasted for nearly ten years and "was begun as a joint project of World Health Organization (WHO), International Labor Organization (ILO), and United Nations Environment Program (UNEP), and later endorsed by the Intergovernmental Forum on Chemical Safety (IFCS). Asia, particularly East Asia and the Pacific islands, was chosen as the feasibility study region . . . [and] the National Institute of Health Sciences (NIHS) of Japan led this initiative and hosted numerous meetings" (Kaminuma 2005; see: http://www.sciencedirect.com/science/article/pii/S0041008X05002772 (accessed August 30, 2017).

5. One can only hope that Diet members with clear policy agendas engage in more critical discussion. This would conform to Leonard Schoppa's (1991) description of "tribes" (*zoku*) of policy experts in the legislature exerting influence collectively on ministry decision-making.

6. This was merely a trend at the time. More concrete action awaited the "Law to Revise Certain Aspects of the Administrative Procedures Act" which came into effect in April of 2006 by Cabinet Order. As James Miller (2006, 16) explains,

> The law requires Japanese regulatory agencies that create new regulatory law to observe a public comment process similar to the model required by the U.S. Administrative Procedures Act (APA). While the law significantly improves the prior cabinet order legal structure, crucial features of the APA and U.S. Regulatory

process are noticeably absent from the adopted Japanese APA. These shortcomings will hamper the realization of the legislation's goals of regulatory transparency and public participation.

7. Although by the end of his Fellowship the Ministry of Health and Welfare (MHW) was already called the "Ministry of Health, Labor, and Welfare," for purposes of this discussion, the older title of the ministry (i.e., MHW) is used in this discussion.

8. Notably, this is an excellent example of the sometimes *reciprocal* nature of the relationship between administrative agents and political principals.

9. Pharmaceutical and Medical Devices Agency, Japan. Visit https://www.pmda.go.jp/english/ for more information (accessed August 30, 2017).

Chapter Six

Case Study 5

Ministry of Economy Trade and Industry (METI)

The final case study considered is the famed Ministry of Economy, Trade, and Industry (METI). Indeed, its predecessor, the Ministry of International Trade and Industry (MITI), has been widely recognized as a key powerbroker in building the modern Japanese Economy and was often trumpeted as the most significant player in the Japanese political system (Johnson 1982). Indeed, not including METI in this discussion might lead to the charge of leaving out potentially one of the most interesting players in the Japanese government, particularly from a historical standpoint. As in the preceding chapters, METI is enmeshed in policy settings involving multiple principals which have created opportunities for bureaucratic drift.

INTRODUCTION

The mission statement of the METI is quite expansive recognizing its continuing role as a powerhouse in shaping the modern Japanese political economy. In 2013, METI boasted that it sought "[t]o respond to every aspect of Japan's economy." As the ministry continues,

> We are looking at various issues from the standpoint of "economic activities." Some divisions look at issues from the industrial standpoint and other divisions look at issues from the cross-industrial standpoint. We implement tight-knit measures, all the warp and woof [sic] of which [sic] is intertwined. In other words, we can support you from a completely new angle. Please talk to our staff members, the like [sic] of which you have never seen before.[1]

Since then, the mission statement has been scaled back to "promoting economic vitality in private companies and smoothly advancing external economic relationships, and to secure stable and efficient supply of energy and mineral resources."[2]

To meet this challenging task, METI's responsibilities are divided between various bureaus with specific responsibilities. For instance, the Economic and Industrial Policy Bureau at which the next Mansfield Fellow was posted (see below), oversees structural reforms and seeks innovative revisions to Japan's infrastructure to facilitate trade.[3] As the bureau's online mission statement explains, this bureau recognizes that to accomplish this goal it must "conduct surveys to obtain data that are not only useful in social terms, but vital to the planning and implementation of its policies" along with "compil[ing] statistics based on these surveys and analyz[ing] the results" in order "to constantly examine, and reform the economic and industrial systems required to assure a bright future for Japan."[4] The bureau itself is then splintered into ten different divisions such as Economic, Industrial Policy Division, the Industrial Structure Policy Division, and the Corporate Affairs Division.

The key point to note is that siloing not only occurs as between different ministries, but also within ministries themselves as different divisions in either the same or different policy bureaus may find themselves with overlapping responsibilities while unable to share findings effectively. For instance, the Trade Policy Bureau has eight different divisions with various responsibilities including five with different geographic responsibilities including the Americas Division, Europe Division, Middle East and Africa Division, Asia and Pacific Division, and the Northeast Asia Division (see below for more on these divisions). The following discussion delves more deeply into the consequences of this fractured bureau and division design when it comes to policy coordination and decision-making. A key term that is employed by a number of Mansfield Fellows to describe METI's internal structure is that of "stovepiping" and their testimony indicated that placement at METI was the most challenging as METI had the hardest time of any other ministry in incorporating the Fellows into their daily work environment. With these preliminary remarks in mind, we turn to the final Mansfield Fellow, Jay Biggs, who was posted at METI.

JAY BIGGS

Mr. Jay Biggs, a Commercial Officer for the U.S. Foreign Commercial Service in the U.S. Department of Commerce, completed his Mansfield Fellowship (2008–2009) in Tokyo serving a six month stint first at the METI studying Japan's industrial promotion strategies and a month in a Diet member's office to see Japan's legislative branch in action. He then spent four months in Japan's Ministry of Health Labor and Welfare (MHLW) studying Japan's healthcare regulatory system, along with one month in the Japanese Cabinet Office of the prime minister to become acquainted with the political system. Biggs holds masters degrees from both the American University in Washington, DC, and from Ritsumeikan University in Kyoto, Japan, and has studied Japanese along with serving for two years in Japan in the U.S. DOC's Office of Japan Trade Policy in Washington, DC, prior to being selected to participate in the Mansfield Fellowship.

As a Mansfield Fellow, Biggs gained extraordinary insight into both the policy network at work within METI along with the larger interplay of interministerial relations in the policy context which included both the Ministry of Foreign Affairs (MOFA) as well as the MHLW. His particular area of policy interest surrounded the regulation of medical devices. However, upon arriving at METI, he discovered that he was to assist with a project comparing the pharmaceutical regulatory framework as between the United States, Japan, and Europe by working with a Japanese administrative counterpart as his mentor at METI. Specifically, his responsibility was to research the U.S. and European systems while his counterpart would investigate the Japanese system in order to present joint findings (Biggs 2010, personal communication).

While Biggs was amenable to this change of plans, he found that the approach by which he and his mentor would proceed was just as challenging as the substantive policy itself. As to the former point, neither he nor his mentor had experience with the pharmaceutical industry. Biggs also discovered that the policy atmosphere was competitive as between different sections in METI itself. Biggs was surprised that his mentor remained "reluctant to talk to other offices within METI" (Biggs 2010, personal communication). Considering the nature of his research and that another office in METI handled pharmaceutical regulatory affairs, Biggs inquired why they were not seeking help from their colleagues who worked in the same building. The response was that it was necessary to wait "until we have something really in place." Consequently, METI struggled for months attempting to garner enough information for the project. To some extent, the bureau-division structure in place compounded the problem. For instance, during this time, Biggs served under two different divisions in the same bureau although his project remained unchanged:

- September–November 2008: Economic and Industrial Policy Bureau, Industrial Structure Policy Division
- December 2008–January 2009: Economic and Industrial Policy Bureau, Americas Division

Likewise, the geographical scope of their project was also complicated by this bureau-division structure. For instance, near the end of his placement in December 2008 through January 2009, he was placed with the Americas Division of the Economic Industrial Policy Bureau although his mentor also expected him to research Europe as well. This mixing of geographic regions added an additional challenge since different divisions were considering the two regions. Two years later, the "Americas Division" was relocated under the "Trade Policy Bureau" rather than the "Economic and Industrial Policy Bureau" and a separate "European Division" also falls under the jurisdiction of that separate bureau as well. The consequence is that the project was more or less doomed from the start as Biggs moved to his third and final stop at METI in February before then continuing on to the Diet in March and the MHLW thereafter. Specifically,

- February 2009: Commerce and Information Policy Bureau, Service Industry Division (Medical and Assistive Device Industries Office)

At the same time, Bigg's mentor was transferred to a different office due to the two-year, cycle rotation system. As such, six months of full-time labor invested in this project were lost as the project was simply abandoned.

This particular case of fractured jurisdictional responsibilities as a consequence of the bureau-division organizational model is not an isolated incident. Other Mansfield Fellowship experiences at various ministries have indicated that seeking advice from offices nearby within the same ministry is a complicated process often requiring traversing multiple levels of hierarchy for weeks to receive permission to walk down the hall to strike up a dialogue with a coordinate office (Biggs 2010, personal communication). As this clarifies, in the context of complex trade regulations, the limitations for penetrating this network by principals is severe absent expertise or some other direct interest link. As such, substantial barriers are created to entry meaning that we would expect administrative discretion to be at its nadir as administrators would be continuously bumping elbows in the protection of their turf. This does indeed track with Biggs's findings. As he explains,

> One of the things social scientists are aware of is that there is a lot of competition between bureaucracies. So, I was aware of that, but I was surprised at just how fierce that was. Not only the battles between MOFA and METI, and MHLW, but even in offices within those bureaucracies (Biggs 2010, personal communication).

To add even more support to this point, Biggs not only observed intra-ministerial competition, but also inter-ministerial jockeying for power. Turning to his final month in February 2009 at METI with medical assisted device industrial regulation, Biggs once again confronted concerns about sharing information (Biggs 2010, personal communication). Medical assisted device regulation was actually an area in which the MHLW was very active. METI's largest concern was that Biggs would soon be posted at MHLW as a Mansfield Fellow and thus might share with MHLW, METI's desire to become involved in this policy area which would result in MHLW's objections and a possible turf war. As Biggs (2010, personal communication) elaborates,

> They were not so concerned about sharing with the U.S. government, but were adamant that I not share anything with MHLW because . . . [t]hey were trying to address how they could possibly tinker with the insurance system giving reimbursement to address some of the cost concerns with the Japanese healthcare system. That is primarily MHLW's issue.

In turn, this also connected with his immediately preceding postings at the Economic and Industrial Policy Bureau's divisions. As Biggs explains,

> So they were essential trying to do . . . research on not only the healthcare system in general, but specifically on the pharmaceutical supply chain and how they could streamline [it]. And the idea was they were going to do some research, they were going to pull together some proposals, and then they were going to announce it in September. And they were going to announce it publicly before they announced it to MHLW in order to get a place at the table on this policy issue. Since I was going to MHLW before that was going to be announced they were really concerned that I not tell MHLW about that (Biggs 2010, personal communication).

This reveals a clear concern for the relative power and bargaining positions of the various ministries in this discrete policy setting. As Biggs succinctly summarizes his experience in this multilayered policy context of medical device analysis, "[s]o that demonstrated the bureaucratic infighting between ministries, between offices within ministries, and so the impact of the two year rotations" (Biggs 2010, personal communication).

This case thus concludes the testing of the primary question under consideration with five different ministries serving as case studies in chapters 2–6 in disparate policy contexts. The Conclusion below summarizes these findings and lends a sense of the way forward for future studies of the Japanese system of public administration. Additionally, the lessons to be learned from this case study about the relationship between administrators and government officials in democratic states is also considered.

CASE STUDY 5: SUMMARY

As this brief final case study of the METI indicates, this policy context implicates relatively few principals in the technically heavy policymaking context. As such, this case confirms that in the absence of multiple principals, bureaucratic discretion is constrained as administrators grow suspicious of one another to the detriment of cooperative policymaking. As all of the foregoing case studies thus confirm, the presence or absence of multiple principals dramatically effects how administrators understand their roles as civil servants.

NOTES

1. See METI's website at http://www.meti.go.jp/english/aboutmeti/data/aOrganizatione/pdf/chart2017.pdf. METI has chronic issues of document translation as revealed in conversations with numerous Fellows who generously and gladly provided document translation assistance during their fellowships with the ministries. Unfortunately, these difficulties are due to METI's need to secure less costly translation services in an era of continuing budget cuts. It would appear that in this instance, cost won over accuracy. According to Merriam-Webster, "warp and woof" is an old English expression dating back to the mid-eighteenth-century meaning "foundation," "base," or "cornerstone" such as "freedom and democracy [are] the warp and woof of any civilized society" (see http://www.merriam-webster.com/dictionary/warp%20and%20woof). The meaning of this expression in this context is thus difficult to decipher. As written, the statement effectively holds that "[w]e implement tight-knit measures, all the foundation and base of which is intertwined." Apparently, the translator seemed to think that "warp and woof" meant "to use a multidimensional approach" to problem solving as the next sentence reveals. When this poor word choice is coupled with the grammatical errors in the final sentence which ought to read the "likes" of "whom" since the first "which" cannot be used to reference people, this should send a minor if clear cautionary note to the Japanese people that administrative agencies cannot be expected to continuously produce the same high quality output with ever decreasing budgets.

2. See METI's website at: http://www.meti.go.jp/english/aboutmeti/data/meti_mission.html (accessed August 31, 2017).

3. See METI's website at: http://www.meti.go.jp/english/aboutmeti/data/aOrganizatione/keizai/keizaiseisaku/01.htm

4. See METI's website at for the bureau's home page: http://www.soumu.go.jp/main_sosiki/gyoukan/kanri/pdf/11_meti.pdf (accessed August 31, 2017).

Conclusions and Areas for Further Study

This study explores the link between democracy and bureaucracy. As this project opined in the opening lines, better understanding of the link between bureaucracy and democracy is critical in modern times as states transition from dictatorships to democracies. Far too often, the focus of inquiry is on the elected branches of government without recognizing that the responsiveness of government cannot be achieved in the presence of significant bureaucratic drift.

Again, *this project asks whether competition by multiple principals creates opportunities for bureaucratic drift?* This text answers affirmatively. Competition by multiple principals *does* often create opportunities for bureaucratic drift, the effect of which is enhanced bureaucratic discretion. Likewise, the absence of multiple principals reduces the opportunities for bureaucratic drift and decreases the exercise of bureaucratic discretion. As this clarifies, variation in agent discretion is the critical effect of the number of principals on bureaucratic drift. Once again, "discretion" is "administrative decision-making absent directives from political principals."

To test the research question, four extended case studies were considered in chapters 2–5 including the Ministry of Internal Affairs and Communications (MIC) (chapter 2), the Ministry of Foreign Affairs (MOFA) (chapter 3), the Ministry of Land, Industry, Transportation, and Tourism (MLITT) (chapter 4), and the Ministry of Health, Labor, and Welfare (MHLW) (chapter 5). Likewise, one brief case study of the Ministry of Economy Trade and Industry (METI) (chapter 6) is included at the end given the interest surrounding this controversial ministry and that it provides yet additional support for the premise that in the absence of multiple principals, fewer avenues for bureaucratic drift are available. Background on each of these ministries is introduced before each of these ministries is assessed and evaluated in various policy contexts. All of the case studies affirm that where multiple principals were involved with their multiplicity of competing interests, opportunities for enhanced bureaucratic drift followed along with the predicted effect of enhanced bureaucratic discretion. Thus, there is significant evidence to support the extension of principal agent theory to Japan, and thus the Asian context more broadly, as political science has done with increasing regularity in recent years (e.g., McCubbins and Noble 1995; Nakano 1998; Ginsburg 2001; Ramseyer and Rasmusen 2001; Miller 2005; Baum 2005; Baum 2007; Yamamoto 2008; Ehrhardt 2009). Moreover, this project finds that Principal Agent Theory provides significant leverage to evaluate policymaking in varied Japanese policy contexts. Indeed, from radio waves to aviation, construction to healthcare, competing multiple principals enhanced opportunities for bureaucratic drift which had the effect of increasing bureaucratic discretion.

Even so, any study must justify itself by answering the deeper question of why this matters? This topic matters because the greater the bureaucratic drift, the more likely agents are to pursue their own policy preferences. While those preferences may align with the expectations of political principals, the case studies adduced above provide ample evidence that this is often not the case. There would be little need to repeatedly seek massive overhauls of the Japanese administrative system unless principals were becoming increasingly irritated (Ehrhardt 2009) with bureaucratic drift. And yet, this project demonstrates that the chosen remedy (i.e., the greater involvement of multiple principals in policymaking), is unlikely to produce the desired effect. Specifically, rather than decreasing bureaucratic drift, involving multiple principals in policymaking is likely to enhance that drift ultimately leaving more discretionary power in the hands of administrators. While the jury is still out, it does indeed seem to be the likely outcome of the most recent civil service reforms of 2014 that have resulted in another layer of bureaucracy, the Cabinet Bureau of Personnel Affairs (CBPA) in derogation of turf originally controlled by the National Personnel Authority (NPA) (Mishima 2017, 10). Regardless of the outcome, further stressing the chains of delegation (Strøm 2000, 267) between principals and agents is undesirable as responsive governance is the likely casualty.

Yet, one might still inquire why Japan is a logical choice to seek answers to these questions? Once again, Japan is an excellent choice to ask fundamental questions about the relationship between bureaucracy and democracy in the hopes of better understanding bureaucratic drift. As clarified in the Introduction, on the one hand, Japan is the only non-western, advanced industrialized democracy. On the other hand, Japan is home to one of the most powerful and controversial administrative systems in the world. Chapter 1 first described the structure of this administrative system evaluating its history, evolution, powers, and roles. One institution in particular, the National

Personnel Authority (NPA), was scrutinized as the agency initially designed by the American Occupiers after World War II to decrease bureaucratic drift in Japan. Indeed, as the premier personnel agency tasked with training successive generations of Japanese administrators, the NPA was intended to serve as the institutional brake on the indiscriminate pursuit of divergent preferences by a technocratic elite. This was to occur through a combination of classic tools some of which are familiar to political control theorists (McCubbins, Noll, and Weingast 1987) and others to public administration efficiency adherents (Mosher 1982). Through carefully designed vetting, training, and ongoing evaluation system of administrators, the NPA was thus intended to ensure that administrators remained "servants of the whole people" under Article 15 of the Constitution.

In practice, however, rather than focusing on reducing bureaucratic drift thus enhancing the responsiveness of administrative agents to political principals, the NPA has devoted increasing attention to enhancing the efficiency of the public administration in Japan. This approach is attuned to the call of public administration theorists that *efficient* administration is the inimitable key to democratic governance. This focus on efficiency is also due to resource constraints as the NPA is understaffed and underfunded, not to mention somewhat demoralized, while seeking to justify its existence to Japanese principals who opposed its establishment by foreign occupiers in the first place. As such, the NPA's "deck-stacking" (Lupia and McCubbins 2000, 302) powers to constrain agents were underused and administrative elites were left relatively unchecked to run the daily affairs of government.

This system went unchallenged during the post-war economic boom years which Chalmers Johnson so aptly summarized as *MITI and the Japanese Miracle* (1982). However, with the financial crisis of the 1990s, the citizenry realized with a start that bureaucratic drift had hit a high water mark (Scheiner and Muramatsu 2009). Thus, starting with strident calls in the 1990s (Ozawa 1994) and launching in 2000, Japan implemented a massive reorganization of the government ministries to realign the goals of bureaucratic agents and political principals. Changing the structure of the agencies themselves it was thought would decrease bureaucratic sectionalism. Iron triangles would thus be shattered as agents realigned in more logical task oriented groups designated by political principals. The outcome was instead heightened and continuing stress on the chains of delegation. Adrift, agents exercised enhanced discretion to pursue their preferences; often in contradistinction to the preferences of political principals and often in poorly merged ministries with disparate policymaking mechanisms and traditions. Continuing public "irritation" (Ehrhardt 2009) was the result.

Given these findings, the imperative question thus becomes, what is the way forward? More broadly, what are the lessons to be learned? As of this writing, Japan's administrative system is still finding its footing in the wake of the 2014 civil service reform as Prime Minister Shinzo Abe seeks to define the state's place particularly in regional security affairs. Whether Japan is prepared to manage the next critical juncture is an open question. The only certainty is that such a critical juncture will be difficult to avoid whether in the form of a tsunami or a rocket as the unpredictable regime of Kim Jong Un's in North Korea aptly demonstrated by firing over Japanese airspace in August of 2017.

To pursue this point for just a moment, Japan's Fukushima nuclear crisis of March 2011 presents a disconcerting case study of how important carefully responding to bureaucratic drift is to the public welfare. In brief, Japan lacks a national disaster management agency along the lines of the U.S. Federal Emergency Management Agency (FEMA) in the United States instead spreading the authority as between the civilian agencies and the Japan Self-Defense Force which is comprised of "special employees" who do not fall under the National Personnel Authority's purview (see table 4.1c) (Bosner 2002; Bosner 2009, personal communication). From a historical standpoint, this has led to "lack of interagency planning or coordination" with the consequence "that Japan's emergency responders will have to cope as best they can in a major disaster with no true comprehensive plan" (Bosner 2002, 20). Japan's present emergency management system, however, was the outgrowth, in part, of the government's poor response to the Kobe earthquake of 1995 which registered 7.2 on the Richter-Scale, killed more than 6,000, injured over 40,000, and left over 300,000 homeless (Bosner 2002, 18). As another Mansfield alumnus (MFP 1999–2001) with expertise in emergency management having traveled to Japan from FEMA (see table 1.1) summarized this earlier disaster,

> The media criticized Japan's government at all levels for what many perceived as a hesitant and uncoordinated response. According to a *Japan Times* special report, in the hours and days after the Kobe earthquake:
>
> 1. The Prefectural Government did not officially request help from the JSDF [Japan Self-Defense Forces] for the first four hours after the quake.
> 2. Lacking specific information, the JSDF initially deployed a relatively small number of troops, and communication failures further delayed the response.
> 3. Offers of help from foreign countries and international aid were turned away.
> 4. The Japan Government's Disaster Management Bureau had a staff of only 36 people to coordinate the disaster response.
> 5. National government agencies lacked procedures for disaster decision-making or coordination, and so each agency went in its own direction, causing confusion (Bosner 2002, 18).

Bosner continues that "[p]refectural and municipal governments also experienced delays due to a lack of organization and planning, and as the delays mounted, damage to life and property increased" (Bosner 2002, 18).

The outcome of the mismanagement of the disaster, among other things, was placing the Disaster Management Bureau in the Cabinet Officer under the Prime Minister and increasing its size (Bosner 2002, 18). As of the time of this writing, the Bureau remains under the Cabinet Office as per the Disaster Countermeasures Basic Act.[1] Thus, the reaction to the Kobe earthquake was to take power away from expert agents and place the fate of the state in the hands of inexpert political principals.

Fast forwarding sixteen years to the March 2011 tsunami reveals eerie parallels. According to the *New York Times*, the severity of the nuclear disaster increased dramatically as a consequence of a poorly functioning policy network headed by Naoto Kan, the Prime Minister of Japan at the time, rather than the ministries (Onishi et al. 2011). Although the facts and details of this case are complex and still in the process of discovery and evaluation, a brief citation from the article will reveal that when power is taken away from agents as the penalty for bureaucratic drift, the result is not necessarily to improve public welfare.

> [T]he . . . prime minister . . . saw the need for quick action but [his] well-founded mistrust of [the] system . . . deprived him of resources he could have used to make better-informed decisions. . . . Mr. Kan struggled to manage the nuclear crisis because he felt he could not rely on the very mechanisms established by his predecessors to respond to such a crisis. . . [C]hoosing instead to rely on a small circle of trusted advisers with little experience in handling a crisis of this scale, blocked him from grasping the severity of the disaster sooner. Sometimes those advisers did not even know all the resources available to them. This includes the existence of a nationwide system of radiation detectors known as the System for Prediction of Environmental Emergency Dose Information, or Speedi. Mr. Terada and other advisers said they did not learn of the system's existence until March 16, five days into the crisis (Onishi et al. 2011).

If this case is any guide, therefore, responding to bureaucratic drift between administrative agents and political principals by simply taking power away from bureaucrats is unwise. Again, what is the lesson? Surely there must be a preferable approach to curtailing bureaucratic drift rather than simply vesting administrative authority in the hands of inexpert principals? Does Japan's nuclear crisis not affirm Weber's tribute to the value of delegating authority to responsive expert administrators?

Yet, apparently, this has not been the lesson drawn from this terrible disaster. Rather, after the disaster, the Japanese administrative state was once again in the crosshairs of a newly irritated public. Specifically, in the face of this national emergency and turmoil, the Diet adopted bills to cut public servant's wages by 7.8 percent annually through 2014 (*Japan Economic Newswire* 2011) and then additional reforms were effectuated. The rationale was telling, "the government plans to allocate some 200 billion yen in expected reduction for the current fiscal year through March 2014 for the reconstruction of areas devastated by the March 11 earthquake and tsunami" (*Japan Economic Newswire* 2011). It is difficult to imagine more direct evidence that the brunt of the calamities which befall Japan are laid on the doorstep of administrators than a statement that public servants are expected to *pay* for a national, natural disaster out of their own pockets.

Perhaps even more striking, amidst this tempest, the bureaucracy has once again been tossed into a surging reformist tide as though fundamentally reworking the administrative state will somehow keep the system afloat. In fact, on the same date that the above pay cuts were announced, legislation to abolish the National Personnel Authority was introduced to the Diet (*Japan Economic Newswire* 2011). This was in tune with past Prime Minister Yukio Hatoyama statement in 2009 that "[w]e must carry out a reform of the civil servant system requiring discussions on whether the National Personnel Authority should continue to exist" (*Yomiuri Shimbun* 2009). To this end, although put off for another year to 2013 (*Japan Times* 2012), the Diet is considering,

> establish[ing] a new agency . . . that will handle affairs related to national servants, including personnel cost management . . . The current National Personnel Authority will be scrapped, according to an outline of a draft plan for civil service reforms to be finalized within March. The new civil service affairs agency will be in charge of devising basic plans on envisioned cuts in personnel costs, managing the size of the workforce at each government agency and institution, and engaging in collective negotiations with national civil servants . . . But the government will stop short of giving civil servants the right to strike for now. It will consider the issue by examining how the new system will work . . . and whether the public will understand the new system (*Jiji Press Ticker Service* 2011).

The final sentence is startling. The notion that fundamental civil service reform rests upon public *understanding* of how lifetime expert administrators carry out the complex functions of government is deeply counterintuitive from a classic Weberian perspective of bureaucracy's "purely technical superiority over any other form of organization." While the author of this work advised the Japanese government at the time that such an outcome would be unwise and fortunately the Cabinet Bureau of Personnel Affairs (CBPA) was apparently the negotiated settlement in 2014, the fact that any such reforms were even on the table before "examining how the new system will work" was stunning. At the time, the NPA was the only entity in the system capable of assisting in this evaluation to ensure any such reforms are carried out in a managed and experienced fashion. Yet, the chains of delegation were so frayed that the ally of the elected officials in this case was in their crosshairs. In short, the torpedo is this time aimed at the personnel system itself perceiving that only by vesting the power to select top level administrators in political hands will the people somehow "prevent sycophancy in the bureaucracy and create an environment in which bureaucrats will gladly help them execute policy initiatives" (*Japan Times* 2010).

Are periodic civil service administrative reforms really the way forward? This project asserts otherwise. Decreasing bureaucratic drift between political principals and administrative agents to enhance public welfare is not accomplished through stripping experts of their power. Targeting the efficiency of a system of public administration through hierarchical reordering does not address the responsiveness of that system. Again, democracy tolerates inefficiency. It cannot, however, tolerate extreme bureaucratic drift. As such, while taking power away from agents and handing it to principals may establish more efficient channels of hierarchy, it does not enhance the responsiveness of the government to the public welfare as needed expertise is thereby sacrificed in the name of "efficiency."

Rather, this project lends support to a third choice and, with regard to Japan, a road less traveled by: reform undertaken *through* rather than *against* the system as designed. This is in tune with political control theory's expectation that chains of delegation may be strengthened through enhancing the deck-stacking oversight powers of political principals. While admittedly resource intensive, this might be accomplished institutionally. In short, why not call upon administrators, in this case the NPA and the CBPA, to serve as operative guarantors of democratic principles? Why not turn these entities into more affirmative "watchdog" bodies of the bureaucracy that the Japanese system is presently without? Rephrased in the language of political control theory, why not refocus the NPA and CBPA on decreasing bureaucratic drift between principals and agents in order to thereby manage bureaucratic discretion? Why not enhance, in other words, the deck-stacking powers of administrators to safeguard the public welfare?

Naturally, calling for enhancing the powers of any group of administrators—even ones intended to be watchdogs—is hardly a popular battle cry in the present political clime in Japan or, likely, anywhere in the world. Certainly the NPA would require a significant boost in personnel and resources at a time when the opposite is the trend for civil service reform. Likewise, it would also mean significant retraining of the bureaucratic ethos away from its laser like focus on efficiency. Yet, what better time to help agents rethink their responsibility to the state than a time of national crisis? The point is that either public servants deserve that title, or they do not. Either they will be trusted partners in democracy building efforts, or they will not. The Japanese case teaches that rather than using critical junctures to further disassemble a besieged system, these moments present opportunities to strengthen the chains of delegation by supervising bureaucratic drift. Rephrased, the recommendation is to anchor a drifting administrative system.

So again, what are the broader lessons to be drawn from this case? Researchers in other advanced democratic states including the United States might inquire whether the Japanese experience provides lessons for rethinking the chains of delegation between democracy and bureaucracy. Might the NPA provide a model to leverage for other states or is it solely the peculiar outgrowth of an occupied state at a moment of weakness? Can institutions be successfully tasked with overseeing deck-stacking to realign the interests of principals and agents thereby reducing bureaucratic drift? As to research projects in other national contexts without experience with democracy, researchers might ask whether and how administrative systems may be designed to account for potential bureaucratic drift rather than merely to achieve hierarchies of efficiency. Again, recognizing that the link between democracy and bureaucracy is rooted not in efficiency, but rather in responsiveness is a lesson that those building democratic states from the ground up should take to heart. Whether new states or those with legacies of effective public administration, the balance that democracies and bureaucracies strike in finding answers to these questions lies at the heart of ensuring the public welfare.

NOTE

1. See http://www.cao.go.jp/en/disaster.html (accessed August 31, 2017).

References

Agnes, Michael and David B. Guralnik. 2002. *Webster's New World College Dictionary*. 4th ed. Cleveland, OH: Wiley Publishing, Inc.

Anderson, James E. 2003. *Public Policymaking*. 5th ed. Boston: Houghton Mifflin Company.

Asahi Shimbun. 2002a. "Mr. Nogami is a Great Man. He has Sacrificed Himself for All" (February 15).

———. 2002b. "Foreign Ministry Bureaucrats Have Much to be Ashamed Of" (April 4).

Asia Pacific Economic Council (APEC). 2010. "The Yokohama Vision—Bogor and Beyond." *The 18th APEC Economic Leader's Meeting*. Yokohama, Japan, November 13–14, 2010. http://www.mofa.go.jp/policy/economy/apec/2010/docs/aelmdeclaration2010_e.pdf (accessed August 30, 2017).

BBC. 2012. "Japan's Ichiro Ozawa Forms Rival Political Party" (July 11). http://www.bbc.co.uk/news/world-asia-18794030 (accessed August 30, 2017).

Baum, Jeeyang. 2005. "Breaking Authoritarian Bonds: The Political Origins of the Taiwan Administrative Procedure Act." *Journal of East Asian Studies* 5: 365–99.

———. April 2007. "Reining in the Bureaucrats: Democratic Transition and Administrative Procedural Reform in Korea." *Governance* 20(2): 233–54.

Beasley, W. G. 1995. *The Rise of Modern Japan: Political, Economic and Social Change since 1850*. New York: St. Martin's Press.

Berg, L., H. Hahn, and J. Schmidhauser. 1976. *Corruption in the American Political System*. Morristown, NJ: General Learning Press.

Biggs, Jay. 2010. "Japan Pharmaceutical Reimbursement." U.S. Department of Commerce.

———. 2010. "Japan: Medical Device Registration." U.S. Department of Commerce.

———. July 23, 2010. Recorded Personal Interview. U.S. Commercial Service, U.S. Department of Commerce, Cleveland, U.S. Export Assistance Center, 600 Superior Avenue, East Suite 700. Cleveland, OH, 44114.

Boling, David. 1999–2001. "Final Report: Mansfield Fellowship." (Private document).

———. Fall 2002. "Legal Reform in Japan: A New Road Map." *Asia Perspectives*: 33–36.

———. Spring 2003. "The Role of Prosecutors in Japanese Antimonopoly Law: Criminal Cases." *Antitrust*: 90–95.

———. September 7, 2009. Recorded telephone interview.

Borthwick, Mark. 2007. *Pacific Century: The Emergence of Modern Pacific Asia*, 3rd ed. Boulder, CO: Westview Press.

Bosner, Leo. August, 2001. "Emergency Management in Tokyo." *Mansfield Report*: 1–4. (Private document).

———. October 2001. "Emergency Management in Tokyo." *Mansfield Report*: 1–18. (Private document).

———. Spring 2002. "U.S.-Asia Security Relations: In the Aftermath of September 11, Emergency Preparedness: A Japan-U.S. Comparison." *Asia Perspectives*: 17–20.
———. September 27, 2009. Recorded telephone interview.
Bozeman, Barry. 2000. *Bureaucracy and Red Tape*. Upper Saddle River, NJ: Prentice Hall.
Bradley, Bill. 2001. "Causes without Rebels: Japanese NGOs Battle Inexperience, Public Apathy" (May 21). *Daily Yomiuri*.
Brehm, John and Scott Gates. 1997. *Working, Shirking, and Sabotage: Bureaucratic Response to a Democratic Public*. Ann Arbor: University of Michigan Press.
Brosnan, Dolores. 1989. "A Comparative Analysis of United States and Japanese Public Personnel Systems." *Asian Journal of Public Administration* 11(1): 1–25.
Cabinet Secretariat, Government of Japan. Administrative Procedure Act. (Act No. 88 of 1993.) http://www.cas.go.jp/jp/seisaku/hourei/data/APA.pdf (accessed August 30, 2017).
Campbell, John Creighton and Ethan Scheiner. 1985. "Review Essay, Fragmentation and Power: Reconceptualizing Policy Making under Japan's 1955 System." *Japanese Journal of Political Science* 9(1): 89–113.
Carothers, Thomas. 2006. *Confronting the Weakest Link: Aiding Political Parties in New Democracies*. Washington, DC: Carnegie Endowment for International Peace.
Carpenter, Susan. 2008. *Why Japan Can't Reform: Inside the System*. New York: Palgrave MacMillan.
Charlton, Sue Ellen M. 2010. *Comparing Asian Politics: India, China, and Japan*. 3rd ed. Boulder, CO: Westview Press.
Chubb, John E. 1985. "The Political Economy of Federalism." *American Political Science Review* 79(4): 994–1015.
Colignon, Richard A. and Chikako Usui. 2003. *Amakudari: The Hidden Fabric of Japan's Economy*. Ithaca, NY, and London: ILR Press an imprint of Cornell University Press.
Cottingham-Streater, Paige. October 22, 2009. Recorded telephone interview.
———. July 19, 2010. "Evaluation of Michael Panzera." Private Letter to Patricia McCarthy, Assistant Director, Civil Division, U.S. Department of Justice. 1100 L Street, N.W. Room 11022, Washington, DC 20530. (Confidential. Copy provided courtesy of Michael Panzera.)
———. January 26, 2011. Personal interview. The Maureen and Mike Mansfield Foundation Tokyo Office, Nakata Mac, Toranomon Building, 1F. 1–1–10 Atago. Minato-ku, Tokyo 105–0002, Japan.
Coy, Patrick G. October, 2001. "Shared Risks and Research Dilemmas on a Peace Brigades International Team in Sri Lanka." *Journal of Contemporary Ethnography* 30(5): 575–606.
Dahl, Robert. 1989. *Democracy and Its Critics*. New Haven, CT: Yale University Press.
della Porta, Donatella and Alberto Vannucci. 1999. *Corrupt Exchanges: Actors, Resources, and Mechanisms of Political Corruption*. New York: Aldine De Gruyter.
Dower, John W. 1986. *War without Mercy: Race and Power in the Pacific War*. New York: Pantheon Books.
———. 1999. *Embracing Defeat: Japan in the Wake of World War II*. New York: W.W. Norton and Company/The New Press.
Ehrhardt, George. 2009. "Administrative Reform in East Asia: The Importance of Multiple Principals." *Asian Survey* 49(4): 625–46.
Ellis, Yukiko. June 15, 2011. Yukiko Ellis: Franklin Lebo. With Comments.
———. July 22, 2009. Yukiko Ellis: Survey Document.
———. July 21, 2009. Recorded telephone interview.
———. November 1, 2004a. Final Report to the Maureen and Mike Mansfield Foundation: 1–10.
———. 2004b. Mansfield Fellowship Monthly Reports for the Period: October 14, 2003–September 4, 2004.
———. 1990. Examination of Census Omission and Erroneous Enumeration Based on 1990 Ethnographic Studies of Census Coverage. *Research Reports and Studies*, U.S. Census Bureau: 0–24.
Esman, Milton J. 1999. Public Administration and Conflict Management in Plural Societies: The Case for Representative Bureaucracy. *Public Administration and Development* 19: 353–66.

Fairbank, John K., Edwin O. Reischauer, and Albert M. Craig. 1989. *East Asia: Tradition and Transformation*. Boston: Houghton Mifflin Company.

Farazmand, Ali (Ed.). 2001. *Handbook of Comparative and Development Public Administration*. New York: Marcel Dekker.

Fenno, Richard. 1990. "The Political Scientist as Participant Observer." In *Watching Politicians: Essays on Participant Observation*, 55–94. Berkeley: IGS Press, Institute of Governmental Studies, University of California at Berkeley.

Fiorina, Morris. 1989. *Congress: Keystone of the Washington Establishment*. 2nd ed. New Haven, CT: Yale University Press.

Fukui, Haruhiro. 1977. "Policymaking in the Japanese Foreign Ministry." In *The Foreign Policy of Modern Japan*, edited by Robert A. Scalapino, 3–36. Berkeley: University of California Press.

Fukumoto, Kentaro. 2011. "Legislators." In *Japanese Politics Today: From Karaoke to Kabuki Democracy*, edited by Takashi Inoguchi and Purnendra Jain, 51–70. New York: Palgrave Macmillan.

Garner, Bryan A. 1996. *Black's Law Dictionary: New Pocket Edition, Revised from A-Z*. Eagan, MN: West Group

Gerlach, Michael L. 1992. *Alliance Capitalism: The Social Organization of Japanese Business*. Berkeley: University of California Press.

Gerring, John. 2001. *Social Science Methodology: A Critical Framework*. Cambridge: Cambridge University Press.

Ginsburg, Tom. Autumn, 2001. "Dismantling the 'Developmental State?' Administrative Procedure Reform in Japan and Korea," *American Journal of Comparative Law* 49(4): 585–625.

Golden, Marissa Martino. 2000. *What Motivates Bureaucrats? Politics and Administration During the Reagan Years*. New York: Columbia University Press.

Goldstein, Kenneth. December, 2002. "Getting in the Door: Sampling and Completing Elite Interviews." *Political Science*: 669–72.

Gormley, William, Jr. Summer 1986. Regulatory Issue Networks in a Federal System. *Polity* 18(4) (Summer): 595–620.

Gormley, William, Jr. and Steven J. Balla. 2008. *Bureaucracy and Democracy: Accountability and Performance*. Washington, DC: Congressional Quarterly Press.

Gross, Ames and Caroline Tran. September 2003. "Recent Revisions in Japan's Pharmaceutical Affairs Law for Medical Devices." *Regulatory Affairs Focus*.

Halloran, Fumiko. 2007. "Kozo Kaikau No Shinjitsu: Takenaka Heizo Daijin Nishi" (Truth of Structural Reforms: Diary of Minister Takenaka Heizo). *Japan Book Review, Japan Society* 2(6): 3–5.

Hancock, M. Donald, David P. Conradt, Guy B. Peters, William Safran, and Raphael Zariski. 1993. *Politics in Western Europe*. Chatham, NJ: Chatham House Publishers, Inc.

Hartcher, Peter. 1998. *The Ministry: How Japan's Most Powerful Institution Endangers World Markets*. Cambridge, MA: Harvard Business School Press.

Hasegawa, Mina. 2000. Merging Matters (December 4). *The Nikkei Weekly (Japan)*.

Havens, Thomas R. H. 1987. *Fire across the Sea: The Vietnam War and Japan, 1965–1975*. Princeton, NJ: Princeton University Press.

Hauss, Charles. 2009. *Comparative Politics: Domestic Responses to Global Challenges*. 6th ed. Australia: Wadsworth Cengage Learning.

Hemphill, Robin R. (Ed.). 2011. "Hypercalcemia in Emergency Medicine." *Medscape Reference*. http://emedicine.medscape.com/article/766373-overview (accessed August 30, 2017).

Hendriks, Carolyn M. 2006. "Integrated Deliberation: Reconciling Civil Society's Dual Role in Deliberative Democracy." *Political Studies* 54: 486–508.

Henkin, Louis. 1972. *Foreign Affairs and the Constitution*. Mineola, NY: Foundation Press.

Hoffman, Susan and Mark Cassel. 2010. *Mission Expansion in the Federal Home Loan Bank System*. Albany: SUNY University Press.

Honig, Bonnie. 2007. Between Decision and Deliberation: Political Paradox in Democratic Theory. *American Political Science Review* 101(1) (February): 1–17.

Hook, Steven W. (Ed.). 2010. *Democratic Peace in Theory and Practice*. Kent, OH: Kent State University Press.

———. 2011. *U.S. Foreign Policy: The Paradox of World Power*. 3rd ed. Washington, DC: Congressional Quarterly Press.

———. 2011. February 3, 2012. "Dissertation Draft." E-mail communication.

Hori, Harumi. 2005. *The Changing Japanese Political System: The Liberal Democratic Party and the Ministry of Finance*. London and New York: Routledge.

Huntington, Samuel P. 1991. *The Third Wave: Democratization in the Late Twentieth Century*. Norman: University of Oklahoma Press.

Hwang, Sung-Don. 1999. *Bureaucracy vs. Democracy in the Minds of Bureaucrats: To What Extent Are These Ideologies Compatible with One Another?* New York: Peter Lang.

Igarashi, Chie. January 26, 2011. Personal interview. The Maureen and Mike Mansfield Foundation Tokyo Office, Nakata Mac, Toranomon Building, 1F. 1–1–10 Atago. Minato-ku, Tokyo 105–0002, Japan.

Iijima, Isao. 2008. "Iijima Isao on the Japanese Political Situation." Ampontan: Japan from the Inside Out." Wordpress.com.http://ampontan.wordpress.com/2008/02/21/iijima-isao-on-the-japanese-political-situation/ (accessed August 30, 2017).

Ikenberry, G. John and Michael Mastanduno (Eds.). 2003. *International Relations Theory and the Asia-Pacific*. New York: Columbia University Press.

Inatsugu, Hiroaki. 2011. "The System of Bureaucrats in Japan." In *Japanese Politics Today: From Karaoke to Kabuki Democracy*, edited by Takashi Inoguchi and Purnendra Jain, 29–49. New York: Palgrave Macmillan.

Inoguchi, Takashi. 2011. "Prime Ministers." In *Japanese Politics Today: From Karaoke to Kabuki Democracy*, edited by Takashi Inoguchi and Purnendra Jain, 11–28. New York: Palgrave Macmillan.

———. 2005. *Japanese Politics: An Introduction*. Melbourne: Trans Pacific Press.

Inoguchi, Takashi and Purnendra Jain (Eds.). 2011. *Japanese Politics Today: From Karaoke to Kabuki Democracy*. New York: Palgrave Macmillan.

International Herald Tribune (Herald Asahi). 2009a. Reform of the Nation's Calcifying Bureaucratic Structure Remains Stalled Despite a Consensus Among Many Lawmakers in Both the Ruling and Opposition Camps that the System Badly Needs to be Revamped (January 14).

———. 2009b. Beijing—Chinese Officials Rolled out the Red Cart Thursday for Veteran Lawmaker Ichiro Ozawa as he Headed a Delegation of 600-Plus Members for a Lightning Visit" (December 12).

ITA (International Trade Administration). December 8, 2000. "Medical Device Regulatory Requirements for Japan: Revision of Japanese Medical Device Approval Applications Classifications," United States International Trade Administration, http://www.ita.doc.gov/td/health/japanrevision.html (accessed August 30, 2017).

Ishimaru, Ken. 2009. Final Report September 2008-September 2009. U.S. Department of Energy, MFP. No. 13: 1–25.

Iitake, Koichi. 1999. MOF to Change Name in Reform Plan, but to Retain Power" (April 16). *Asahi Shimbun (Asahi Evening News), Pg. News.*

Janda, Kenneth, Berry, Jeffrey M., Goldman, Jerry. 1997. *The Challenge of Democracy: Government in America*. 5th ed. Boston: Houghton Mifflin Company.

Janis, Irving L. and Leon Mann. 1977. *Decision Making: A Psychological Analysis of Conflict, Choice, and Commitment*. New York: Free Press.

Japan Economic Newswire. 2011. "Government Eyes Giving Local Public Servants Right to Bargain over Wages." (June 6).

Japan Times. 2010. "Central Control over Bureaucrats." (February 26). http://www.japantimes.co.jp/text/ed20100226a1.html (accessed August 30, 2017).

———. 2012. "Administrative Reform Bills Likely to Languish: Civil Service Put on Hold as Noda Pushes to Pass Sales Tax Hike." (July 31).

Jiji Press Ticker Service. 2011. "Japan Eyes New Agency for National Civil Service." (March 1).

Johnson, Chalmers. 1982. *MITI and the Japanese Miracle: The Growth of Industrial Policy, 1925–1975*. Stanford, CA: Stanford University Press.

———. 1995. *Japan: Who Governs?—The Rise of the Developmental State*. New York: W.W. Norton and Company.

Journal of Japanese Trade and Industry (JJTI). 2001. "Former Ministry Official Arrested for Swindling Secret Government Fund" (May 1).

Kamada, Kazuko. January 27, 2011. Personal Interview. International Affairs Division, Secretariat of the National Personnel Authority. 1–2–3 Kasumigaseki, Chiyoda-ku,Tokyo, 100–8913, Japan.

Kaminuma, Tsuguchika. September 2005. "Lessons Learned in Building a Global Information Network on Chemicals (GINC)." *Toxicology and Applied Pharmacology* 207(2): 750–54. http://www.sciencedirect.com/science/article/pii/S0041008X05002772 (accessed August 30, 2017).

Kaneko, Yuko and Itoko Suzuki. 2007. "Recent Major Administrative Reforms: Japan's Response to Global and Domestic Challenges." In *Handbook of Globalization, Governance, and Public Administration*, edited by Ali Farazmand and Jack Pinkowski, 519–52. Boca Raton, FL; London; and New York: Taylor and Francis Group.

Kariya, James P. January, 1998. "Final Report." *Mansfield Fellowship Program*, 1–12.

———. February 23, 2009. E-mail communication.

Kaufman, Herbert. 1960. *The Forest Ranger: A Study in Administrative Behavior*. Baltimore: Johns Hopkins University Press.

Kawai, Masahiro and Shinji Takagi. 2004. "Japan's Official Development Assistance: Recent Issues and Future Directions." *Journal of International Development* 16: 255–80.

King, David C. 1997. *Turf Wars: How Congressional Committees Claim Jurisdiction*. Chicago and London: University of Chicago Press.

Kobayashi, Ken. 2000a. Monthly Report to FDA: November 1, 2000.

———. 2000b. Monthly Report to FDA: December 1, 2000.

———. 2001. Monthly Report to FDA: January 1, 2001.

———. December 8, 2010. Recorded telephone interview.

Koh, B.C. 1989. *Japan's Administrative Elite*. Berkeley: University of California Press.

Kohno, Masaru. 1997. *Japan's Postwar Party Politics*. Princeton, NJ: Princeton University Press

Krauss, Ellis S. and Bejamin Nyblade. 2005. "'Presidentialization' in Japan? The Prime Minister, Media, and Elections in Japan." *British Journal of Political Science* 35: 357–68.

Krauss, Ellis S. and Robert J. Pekkanan. 2011. *The Rise and Fall of Japan's LDP*. Ithaca, NY: Cornell University Press.

Kuramoto, Teruyo. January 27, 2011. Personal interview. The Maureen and Mike Mansfield Foundation Tokyo Office, Nakata Mac, Toranomon Building, 1F. 1–1–0 Atago. Minato-ku, Tokyo 105–0002, Japan.

Lane, Frederick S. (Ed.). 1994. *Current Issues in Public Administration*. 5th ed. New York: St. Martin's Press.

Laxmikanth, M. 2009. *Public Administration: For the UPSC and State Civil Services Preliminary Examinations*. New Delhi: Tata McGraw Hill.

Lebo, Franklin Barr. 1999. "Bureaucratic Control of Japan: The Development and Functioning of a Traditional System." Senior Honors Thesis. Defended at Brandeis University on December 10, 1999. Copyrighted and on reserve in the Robert D. Farber University Archives and Special Collections, Brandeis University, 1–110.

Lijphart, Arend. 1984. *Democracies: Patterns of Majoritarian and Consensus Government in Twenty-One Countries*. New Haven, CT: Yale University Press.

Lu, David. 1997. *Japan: A Documentary History—The Late Tokugawa Period to the Present*. London and New York: Routledge.

Lupia, Arthur and Mathew McCubbins. (2000). "Representation or Abdication? How Citizens Use Institutions to Help Delegation Succeed." *European Journal of Political Research* 37: 291–307.

Maclachlan, Patricia L. 2009. "Two Steps Forward, One Step Back: Japanese Postal Privatization as a Window on Political and Policymaking Change." In *Political Change in Japan:*

Electoral Behavior, Party Realignment, and the Koizumi Reforms, edited by Steven R. Reed, Kenneth Mori McElwain, and Kay Shimizu, 157–77. Stanford, CA: The Walter H. Shorenstein Asia-Pacific Research Center.

Maeda, Ko. 2009. "Has the Electoral-System Reform Made Japanese Elections Party-Centered?" In *Political Change in Japan: Electoral Behavior, Party Realignment, and the Koizumi Reforms*, edited by Steven R. Reed, Kenneth Mori McElwain, and Kay Shimizu, 47–66. Stanford, CA: The Walter H. Shorenstein Asia-Pacific Research Center.

Maier, Brent. 1999. Mansfield Fellow Final Report: September 1997-September 1999. U.S. Environmental Protection Agency, San Francisco Regional Office. (Private document).

———. August 10, 2009. Recorded telephone interview.

Mainichi Daily News. 2003. "Pen-Pushers Commit 17,000 Unreported Offenses" (October 6).

March, J.G. and J.P. Olsen. 1989. *Rediscovering Institutions*. New York: Free Press.

Marcus, Michael J. December 30, 1999. "Final Report of Michael J. Marcus" (edited version). *Federal Communications Commission.*

Marshall, Gordon. 1998. "Ideographic versus Nomothetic Approaches." *A Dictionary of Sociology*. www.encyclopedia.com (accessed August 30, 2017).

Martin, Sherry L. and Gill Steel (Eds.). 2009. Democratic Reform in Japan: Assessing the Impact. Boulder, CO: Lynne Rienner Publishers.

Matsumura, Shigeo. 1997. "The Strong and Controversial Finance Ministry: Beyond the Bounds of Propriety" (June 3). *Asahi Shimbum (Asahi Evening News)*. Pg. News.

McCraw, Thomas K. 1984. Prophets of Regulation: Charles Francis Adams, Louis D. Brandeis, James M. Landis, Alfred E. Kahn. Cambridge, MA: The Belknap Press of Harvard University Press.

McCubbins, Matthew and Gregory Noble. 1995. "The Appearance of Power: Legislators, Bureaucrats, and the Budget Process in the United States and Japan." In Structure and Policy in Japan and the United States, edited by P. F. Cowhey and M. D. McCubbins. Cambridge: Cambridge University Press.

McCubbins, Matthew D., Roger G. Noll, and Barry R. Weingast. 1987. "Administrative Procedures as Instruments of Political Control." *Journal of Law, Economics, and Organization* 3(2): 243–77.

———. March 1989. "Structure and Process, Politics and Policy: Administrative Arrangements and the Political Control of Agencies." *Virginia Law Review* 75(2): 431–82.

McKinney, Lucinda. 1999. "Technical Complexity, Regionalism, and Decision-Making: Case Processing Outcomes in EPA's Toxic Substances Program." The Social Science Journal, 35(3): 377–92.

Meier, Kenneth J. and Laurence J. O'Toole Jr. 2006. Bureaucracy in a Democratic State: A Governance Perspective. Baltimore: The Johns Hopkins University Press.

Meyer, Steven A. and Shigeto Naka. March 1998. "Legislative Influences in Japanese Budgetary Politics." Public Choice 94(3–4): 267–88.

McElwain, Kenneth Mori and Steven R. Reed. 2009. "Japanese Politics in the Koizumi Era: Temporary Anomaly or a Paradigm Shift?" In *Political Change in Japan: Electoral Behavior, Party Realignment, and the Koizumi Reforms*, edited by Steven R. Reed, Kenneth Mori McElwain, and Kay Shimizu, 281–91. Stanford, CA: The Walter H. Shorenstein Asia-Pacific Research Center.

Miller, Gary. 2005. "The Political Evolution of Principal-Agent Models." *Annual Review of Political Science* 8: 203–25.

Miller, James J. October 30, 2006. Mike Mansfield Fellowship Program: Final Report of James J. Miller. Federal Communications Commission. Mansfield Fellowship Program #10. September 2004–September 2006.

Ministry of Health, Labor, and Welfare (MHLW). 2007. "Chapter 9, Section 2: Promotion of Measures to Provide Better Drugs Safely and Promptly." *Annual Health, Labour, and Welfare Report of 2007–2008* (unofficial translation). http://www.mhlw.go.jp/english/wp/wp-hw2/index.html (accessed August 30, 2017).

———. March 2010a. Ministry of Health, Labour, and Welfare: Pamphlet. Chuo Godo Chosha 5 go-kan, 1–2–2 Kasumigaseki, Chiyoda-ku, Tokyo 100–8916, pp. 1–35.http://www.mhlw.go.jp/english/org/pamphlet/dl/pamphlet-about_mhlw.pdf(accessed August 30, 2017).

———. 2010b. "Section III: Internal Reform Efforts." *Annual Health, Labour, and Welfare Reports (2009–2010)*. http://www.mhlw.go.jp/english/wp/wp-hw4/honbun.html (accessed August 30, 2017).

Ministry of Internal Affairs and Communications (MIC). 2010. White Paper: Information and Communications in Japan. Economic Research Office, ICT Strategy Policy Division, Global ICT Strategy Bureau, MIC, Japan. 1–2, Kaumigaseki 2-chome, Chiyoda-ku Tokyo 100–6926, Japan. http://www.soumu.go.jp/johotsusintokei/whitepaper/eng/WP2010/2010-index.html (accessed August 30, 2017).

Ministry of Land, Infrastructure, and Transport (MLIT). 2001. White Paper on Land, Infrastructure, and Transport in Japan, 2001.

Mishima, Ko. January, 2017. "A Big Bang for Japanese Mandarins? The Civil Service Reform of 2014." *International Journal of Public Administration*, online: 1–13. http://www.tandfonline.com/doi/abs/10.1080/01900692.2016.1242615?journalCode=lpad20 (accessed August 30, 2017).

Mitchell, David and Tansa George Massoud. July, 2009. Anatomy of Failure: Bush's Decision-Making Process and the Iraq War. *Foreign Policy Analysis* 5(3): 265–86.

Moe, Terry M. 1989. "The Politics of Bureaucratic Structure." In *Can the Government Govern?* edited by John E. Chubb and Paul E. Peterson, 267–329. Washington, DC: Brookings Institution.

Mosher, Frederick C. 1982. *Democracy and the Public Service*. 2nd ed. New York and Oxford: Oxford University Press.

Mueller, John. 1999. *Capitalism, Democracy, and Ralph's Pretty Good Grocery*. Princeton, NJ: Princeton University Press.

———. 2010. "Faulty Correlation, Foolish Consistency, Fatal Consequence." In *Democratic Peace in Theory and Practice*, edited by Steven W. Hook, 45–60. Kent, OH: Kent State University Press, 45–60.

Mulgan, Aurelia George. 2003. "Japan's Un-Westminster System: Impediments to Reform in a Crisis Economy." In *Government and Opposition*. Oxford: Blackwell Publishing.

———. 2011a. "The Politics of Economic Reform." In *The Routledge Handbook of Japanese Politics*, edited by Alisa Gaunder, 261–72. New York: Routledge.

———. 2011b. "The Farm Lobby." In *Japanese Politics Today: From Karaoke to Kabuki Democracy*, edited by Takashi Inoguchi and Purnendra Jain, 109–26. New York: Palgrave Macmillan.

Murayama, Yuji. 2000. *Japanese Urban System*. Dordrecht; Boston; and London: Kluwer Academic Publishers.

Nachmias, David and David H. Rosenbloom. 1980. *Bureaucratic Government USA*. New York: St. Martin's Press.

Nagasaka, Satoru, Benjamin Lang, Mihoko Shintani, and Sayaka Ueno. April, 2008. An Overview of Pharmaceutical and Medical Device Regulation in Japan. *Special Report: Life Sciences, Asian Council*. Morgan Lewis-TMI: 28–35.

Nakamura, Akira. 2005. "The Debilitating Power of Japan's Central Government Bureaucrats: Reality or Fallacy?" In *Public Service Reform in East Asia: Reform Issues and Challenges in Japan, Korea, Singapore, and Hong Kong*, edited by Anthony B. L. Cheung, 19–38. Hong Kong: The Chinese University Press.

———. 2012. "Asian Model of Government Re-examined in the Aftermath of the Global Economic Crunch: A Japanese Perspective from the Experience of the Triple Disasters in March 2011." *International Review of Administrative Science* 78(2): 239–59.

Nakano, Koichi. 1998. "The Politics of Administrative Reform in Japan, 1993–1998." *Asian Survey* 38(3): 291–309.

National Personnel Authority. October 1, 2000. *NPA Newsletter*. Vol. 1. International Affairs Division, National Personnel Authority, Government of Japan.

———. 2002. *NPA Newsletter: Recent Movements of NPA and Personnel Administration in Japan*. International Affairs Division, National Personnel Authority, Government of Japan.

———. 2005. *Annual Report: FY 2004 (April 2004–March 2005)*. Published by the International Affairs Division of the National Personnel Authority, Government of Japan. 1–2–3 Kasumigaseki, Chiyoda-ku, Tokyo 100–8913, JAPAN. (English edition.)

———. January 2006. *NPA Newsletter*. Vol. 6. International Affairs Division, National Personnel Authority, Government of Japan.

———. 2006. *Annual Report: FY 2005 (April 2005–March 2006)*. Published by the International Affairs Division of the National Personnel Authority, Government of Japan. 1–2–3 Kasumigaseki, Chiyoda-ku, Tokyo 100–8913, JAPAN. (English edition.)

———. 2007. *Annual Report: FY 2006 (April 2006–March 2007)*. Published by the International Affairs Division of the National Personnel Authority, Government of Japan. 1–2–3 Kasumigaseki, Chiyoda-ku, Tokyo 100–8913, JAPAN. (English edition.)

———. 2008. *Annual Report: FY 2007 (April 2007–March 2008)*. Published by the International Affairs Division of the National Personnel Authority, Government of Japan. 1–2–3 Kasumigaseki, Chiyoda-ku, Tokyo 100–8913, JAPAN. (English edition.)

———. March 2008. *NPA Newsletter*. Vol. 8. International Affairs Division, National Personnel Authority, Government of Japan.

———. March 2009. *NPA Newsletter*. Vol. 9. International Affairs Division, National Personnel Authority, Government of Japan.

———. June 2009. 2009 Monitor of National Civil Servants: Survey Results. Unofficial English translation courtesy of Professor Fredrick Butcher.

———. 2010a. *Annual Report: FY 2008 (April 2008-March 2009)*. Published by the International Affairs Division of the National Personnel Authority, Government of Japan. 1–2–3 Kasumigaseki, Chiyoda-ku, Tokyo 100–8913, JAPAN. (English edition.)

———. 2010b. *2010 Profile of National Public Employees*. Published by the General Affairs Division, General Secretariat of the National Personnel Authority, Government of Japan. 1–2–3 Kasumigaseki, Chiyoda-ku, Tokyo 100–8913, JAPAN. (English edition.)

———. 2010c. "Training and Human Resource Development." Booklet created by the Training and Development Policy Division, National Personnel Authority, Government of Japan. 1–2–3 Kasumigaseki, Chiyoda-ku, Tokyo 100–8913, JAPAN: 1–7.

———. 2016. *2016 Profile of National Public Employees*. Published by the General Affairs Division, General Secretariat of the National Personnel Authority, Government of Japan. 1–2–3 Kasumigaseki, Chiyoda-ku, Tokyo 100–8913, JAPAN. (English edition.) http://www.jinji.go.jp/en/infonpa/profile.pdf (accessed August 30, 2017).

Neary, Ian. 2002. *The State and Politics in Japan*. Cambridge, UK: Polity Press.

Nishio, Takashi et al. February 2010. "Cultivating Professional Public Employees in the New Age—Image of 'the Servants of the Whole People' that Support the Political Leadership." Study Group on Public Service Training and Human Resource Development. Research Society on the Development of Human Resources in coordination with the National Personnel Authority. (Japanese version courtesy of NPA.) Unofficial translation by Professor Fredrick Butcher.

Noble, Gregory. 2011. "The Evolution of the Japanese Policymaking System." In *The Routledge Handbook of Japanese Politics*, edited by Alisa Gaunder, 249–60. New York: Routledge.

Noguchi, Yukio. 1995. "The Role of the Fiscal Investment Loan Program in Postwar Japanese Economic Growth." In *The Japanese Civil Service and Economic Development*, edited by Hyung-ki Kim et al., 261–87. New York: Clarendon Press.

Nye, Joseph S. 1967. "Corruption and Political Development: A Cost-Benefit Analysis." *American Political Science Review* 61(2): 417–27.

Odendahl, Teresa and Aileen M. Shaw. 2002. "Interviewing Elites." In *Handbook of Interview Research*, edited by Jaber F. Gubrium and James A. Holstein, 299–316. Thousand Oaks, CA; London; and New Delhi: Sage Publications.

Okada, Yosuke and Takahito Kushi. July 2004. "Government-Sponsored Cooperative Research in Japan: A Case Study of the Organization of Pharmaceutical Safety and Research (OPSR) Program." OPIR Research Paper Series No. 22. Office of Pharmaceutical Industry Research.

Onishi, Norimitsu, Martin Fackler, and Kantaru Suzuki. June 13, 2011. "In Nuclear Crisis, Crippling, Mistrust." *New York Times*, Section 1, Column 0, Foreign Desk, Pg.1.

Oyama, Reiko. June 30, 2015. "The Rightful Role of the House of Councillors." *Nippon.com: Your Doorway to Japan*. http://www.nippon.com/en/in-depth/a04402/ (Accessed November 23, 2017).

Ozawa, Ichiro. 1994. *Blueprint for a New Japan: The Rethinking of a Nation*. Tokyo: Kodansha International.

Panzera, Michael. 2008–2009. "Monthly Progress Reports." *Mansfield Reports*. (Private documents.)

———. February 6, 2011. Recorded telephone interview.

———. June 8, 2011. E-mail communication.

Peters, Guy B. 1999. *Institutional Theory in Political Science: The "New Institutionalism."* London and New York: Continuum.

Preston, Thomas and Paul t'Hart. 1999. "Understanding and Evaluating Bureaucratic Politics: The Nexus between Political Leaders and Advisory Systems." *Political Psychology* 20(1): 49–98.

Rao, Naveen C. 2003–2005. "Monthly Progress Reports." *Mansfield Reports*: 1–11. (Private document).

———. September 15, 2009. Recorded telephone interview.

Ramseyer, Mark and Eric B. Rasmusen. June 2001. "Why Are Japanese Judges So Conservative in Politically Charged Cases?" *American Political Science Review* 95(2): 331–44.

Ramseyer, J. Mark and Rosenbluth, Frances M. 1993. *Japan's Political Marketplace*. Cambridge, MA: Harvard University Press.

Reed, Steven R. 1993. *Making Common Sense of Japan*. Pittsburgh and London: University of Pittsburgh Press.

———. 2011. "Winning Elections in Japan's New Electoral Environment." In *Japanese Politics Today: From Karaoke to Kabuki Democracy*, edited by Takashi Inoguchi and Purnendra Jain, 71–87. New York: Palgrave Macmillan.

Reed, Steven R., Kenneth Mori McElwain, and Kay Shimizu (Eds.). 2009. *Political Change in Japan: Electoral Behavior, Party Realignment, and the Koizumi Reforms*. Stanford, CA: The Walter H. Shorenstein Asia-Pacific Research Center.

Reed, Steven R. and Kay Shimizu. 2009. "An Overview of Postwar Japanese Politics." In *Political Change in Japan: Electoral Behavior, Party Realignment, and the Koizumi Reforms*, edited by Steven R. Reed, Kenneth Mori McElwain, and Kay Shimizu, 5–46. Stanford, CA: The Walter H. Shorenstein Asia-Pacific Research Center.

Rogow, Arnold and Harold Lasswell. 1966. *Power, Corruption and Rectitude*. Englewood Cliffs, NJ: Prentice Hall

Rondinelli, Dennis A. and John D. Montgomery. 2005. "Regime Change and Nation Building: Can Donors Restore Governance in Post-Conflict States?" *Public Administration and Development* 25: 15–23.

Rosenbaum, Walter A. 2011. *Environmental Politics and Policy*. 8th ed. Washington, DC: CQ Press.

Rosenbluth, Frances M. 1989. *Financial Politics in Contemporary Japan*. Ithaca, NY: Cornell University Press.

Rourke, Francis E. 1984. *Bureaucracy, Politics, and Public Policy*. Boston: Little Brown and Company.

Rubin, Herbert and Irene Rubin. 1995. *Qualitative Interviewing: The Art of Hearing Data*. Thousand Oaks, CA: Sage.

Sabatier, Paul A. (Ed.). 1999. *Theories of the Policy Process*. Boulder, CO: Westview Press.

Saltzburg, Stephen A., John L. Diamond, Kit Kinports, and Thomas H. Morawetz. 2000. *Criminal Law*. 2nd ed. New York: Matthew Bender and Company.

Sato, Chihiro. August 28, 2007. "Government May Set up Tourism Agency: New Entity Would Be Tasked with Improving Japan's Rank among Top Nations." *Daily Yomiuri (Tokyo)*, pg. 4.

Scalapino, Robert (Ed.). 1977. *The Foreign Policy of Modern Japan*. Berkeley: University of California Press.

Scheiner, Ethan and Michio Muramatsu. 2009. "The Slow Government Response to Japan's Bank Crisis: A New Interpretation." In *Political Change in Japan: Electoral Behavior, Party Realignment, and the Koizumi Reforms*, edited by Steven R. Reed, Kenneth Mori McElwain, and Kay Shimizu, 181–98. Stanford, CA: The Walter H. Shorenstein Asia-Pacific Research Center.

Scheyvens, Henry. 2005. "Reform of Japan's Official Development Assistance: A Complete Overhaul or Merely a Fresh Coat of Paint?" *Progress in Development Studies* 5(2): 89–98.

Schlesinger, Jacob M. 1997. *Shadow Shoguns: The Rise and Fall of Japan's Postwar Political Machine*. New York: Simon and Schuster.

Schoppa, Leonard. 1991. "Zoku Power and LDP Power: A Case Study of the Zoku Role in Education Policy." *Journal of Japanese Studies* 17: 79–106.

Schrecker, John. 1991. *The Chinese Revolution in Historical Perspective*. New York: Praeger.

Seawright, Jason and John Gerring. 2008. "Case Selection Techniques in Case Study Research." *Political Research Quarterly* 61(2): 294–308.

Seay, Jeffrey. 1999–2000. "The Mike Mansfield Fellowship—Final Report." *Mansfield Report*: 1–14. (Private document).

Shambaugh, David and Michael Yahuda (Eds.). 2008. *International Relations of Asia*. Lanham, MD: Rowman & Littlefield.

Sheridan, Kyoko. 1993. *Governing the Japanese Economy*. Malden, MA: Polity Press.

Shinoda, Tomohito. 2011. "Prime Ministerial Leadership." In *The Routledge Handbook of Japanese Politics*, edited by Alisa Gaunder, 48–59. New York: Routledge.

Stanley-Baker, Joan. 1984. *Japanese Art*. Singapore: Thames and Hudson.

Stockwin, J.A.A. 2011a. "The Rationale for Coalition Government." In *The Routledge Handbook of Japanese Politics*, edited by Alisa Gaunder, 261–72. New York: Routledge.

———. 2011b. "Party Politics in Japan." In *Japanese Politics Today: From Karaoke to Kabuki Democracy*, edited by Takashi Inoguchi and Purnendra Jain, 89–107. New York: Palgrave Macmillan.

———. 1999. *Governing Japan: Divided Politics in a Major Economy*. Oxford: Blackwell Publishers Ltd.

Strøm, Kaare. 2000. "Delegation and Accountability in Parliamentary Democracies." *European Journal of Political Research* 37: 261–98.

Swink, Larry. July 24, 2009. E-mail communication.

Suleiman, Ezra. 2003. *Dismantling Democratic States*. Princeton and Oxford: Princeton University Press.

Suzuki, Kenji. 2006a. "Policy Network for Network Policy in Japan." In *Japanese Telecommunications: Market and Policy in Transition*, edited by Ruth Taplin and Masako Wakui, 182–205. London and New York: Routledge.

———. 2006b. "The Changing Pattern of *Amakudari* Appointments: The Case of Regional Banks, 1991–2000." In *Institutional Change in Japan*, edited by Magnus Blomstrom and Sumner La Croix, 202–20. London and New York: Routledge.

Takayasu, Kensuke. 2005. "Prime-Ministerial Power in Japan: A Re-Examination." *Japan Forum* 17(1): 163–84.

Tanikawa, Miki. 2017. "Japan's Teflon Premier: Regardless of Scandals or Sluggish Economy, the Public Supports Its Prime Minister" *U.S. News*. (June 1). https://www.usnews.com/news/best-countries/articles/2017-06-01/japans-shinzo-abe-remains-comfortably-in-power (accessed August 30, 2017).

Taplin, Ruth and Masako Wakui (Eds.). 2006. *Japanese Telecommunications: Market and Policy in Transition*, London and New York: Routledge.

The Daily Yomiuri (Tokyo). 2000a. "Expert Panel to Discuss Nuclear Nonproliferation" (February 11), Pg. 2.

———. 2000b. "International Experts Meet to Discuss Nuclear Nonproliferation" (February 25), pg. 2.

———. 2001 (Sunday). "Ministry Saw 146 Discipline Cases in 4-yr Span" (May 13), pg. 2.

———. 2002 (Sunday). "Cliques Hinder Merged Ministries" (January 6), pg. 3.

———. 2004. "Reform Buds in Bureaucrat Personnel System" (July 6), pg. 8.

———. 2008a. "Pub Taxi Issue Reveals Work Disparities" (June 30), pg. 3.

———. 2008b. "Japan Tourism Agency Inaugurated" (October 2), pg. 3.

Turner, Bryan S. 1989. "Research Note: From Orientalism to Global Sociology." *Sociology* 23(4) (November): 629–38

U.S. Office of Personnel Management (OPM). 2011. "The Beginning of Modern Personnel Administration: 1938–1958." 1900 E Street NW, Washington, DC, 1–33. http://www.opm.gov/BiographyofAnIdeal/SubMain1938–1958.asp (accessed August 30, 2017).

Waldo, Dwight. 1980. *The Enterprise of Public Administration: A Summary View*. Novato, CA: Chandler and Sharp Publishers, Inc.

Waller, Anthony J. 2009. 2008–2009 Final Fellowship Report: Mansfield Fellowship Training.

———. Recorded telephone interview. July 16, 2010.

———. June 8, 2011. Bio: Anthony J. Waller. (Private document).

Watanabe, Tadakazu. January 27, 2011. Personal interview. International Affairs Division, Secretariat of the National Personnel Authority. 1–2–3 Kasumigaseki, Chiyoda-ku, Tokyo, 100–8913, Japan.

Weber, Max. 1969. "The Three Types of Legitimate Rule." In *A Sociological Reader on Complex Organizations*. 2nd ed., edited by Amitai Etzioni, 6–14. New York: Holt.

———.2003 (1946). "Bureaucracy." In *Representative Bureaucracy: Classic Readings and Continuing Controversies*, edited by Julie Dolan and David H. Rosenbloom, 8–11. New York and London: M. E. Sharpe.

Williams, Justin. June 1968. "Completing Japan's Political Reorientation, 1947–1952: Crucial Phase of the Allied Occupation." *The American Historical Review* 73(5): 1454–69.

Wilson, Woodrow. 2005 (1887). "The Study of Administration." In *Woodrow Wilson: The Essential Political Writings*, edited by Ronald J. Pestritto, 231–50. Lanham, MD: Lexington Books.

Winter, Markus. 2016. "Abe and the Bureaucracy: Tightening the Reins." *The Diplomat*. (June 16). http://thediplomat.com/2016/06/abe-and-the-bureacracy-tightening-the-reins/ (accessed August 30, 2017.)

Wright, Maurice. 2002. *Japan's Fiscal Crisis: The Ministry of Finance and the Politics of Public Spending, 1975–2000*. Oxford and New York: Oxford University Press.

Yamamoto, Kiyoshi. April 2008. "Has Agencification Succeeded in Public Sector Reform? Realities and Rhetoric in the Case of Japan." *Asian Journal of Political Science* 16(1): 24–40.

Yates, Douglas. 1982. *Bureaucratic Democracy: The Search for Democracy and Efficiency in American Government*. Cambridge, MA: Harvard University Press.

Yetiv, Steve. 2011. *Explaining Foreign Policy: U.S. Decision-Making in the Gulf Wars*. 2nd ed. Baltimore: Johns Hopkins University Press.

Yomiuri Shimbun. 2008. "Editorial: Can Personnel Agency Function Adequately?" (April 5).

———. 2009. "Editorial: Hiring Right Bureaucrat for Right Job Is No TA" (November 19).

———. 2011. "Twists ahead on Road to Civil Service Reform" (January 5).

Yonemaru, Tsuneharu. 2004. "Electronic Government in Japan." In *National Electronic Government: Comparing Governance Structures in Multi-Layer Administrations*, edited by Martin Eifert and Jan Ole Puschel, 136–81. London and New York: Routledge, Taylor U, and Francis Group.

Zegart, Amy. 1999. *Flawed by Design: The Evolution of the CIA, JCS, and NSC*. Stanford, CA: Stanford University Press.

Index

About the Author

Franklin Barr Lebo is assistant professor of sustainability and an Asian Studies Program faculty member at Baldwin Wallace University. Growing up in San Francisco, California, and serving as a host brother to scores of foreign exchange students, he developed a keen interest in Asia and Japan, in particular. Franklin went on to graduate summa cum laude, Phi Beta Kappa, and Pi Sigma Alpha from Brandeis University with a BA in political science and East Asian studies studying Japanese both at Brandies and at the Showa Daigaku Boston Institute. He was a finalist in the Japanese speech competition hosted by the Japan Society of Boston. He then earned a JD at the University of California, Hastings College of the Law with an International Law Concentration and won the CALI Award for his research on Chinese Trade and Investment. Subsequently, he earned a PhD at Kent State University focusing on transnational and comparative politics and policy with a secondary concentration in justice studies. He was an invited member of the Young Scholars Delegation to Taiwan and completed in situ research on the National Personnel Authority in the Kasumigaseki district of Tokyo, Japan. Some of his recent publications include "Proposing a New Framework: Including China on the Crossroads of Transitional Justice and Reconciliation" in the *Asian Journal of Comparative Politics* (2016) and "Sino-American Trade Relations: Privatizing Foreign Policy" with Steven Hook in *Contemporary Cases in U.S. Foreign Policy: From Terrorism to Trade*, 5th Edition (2014). His continuing research interests center on East Asia and sustainable energy.

www.ingramcontent.com/pod-product-compliance
Lightning Source LLC
LaVergne TN
LVHW011520200926
848484LV00001B/85

* 9 7 8 1 4 9 8 5 6 2 2 1 8 *